PRAISE FOR MS.

"Dorinne Davis has excavated the auditory field with amazing discoveries in the world of ear, voice, brain, and language. She brings to many fields the gift of insight, wisdom, and clinical design that can help parents, teachers, and researchers move forward with brilliant perspectives."

Don Campbell, author of *The Mozart Effect*

"Dorinne Davis is one of the leading edges in the field of modern sound work. Her research, commitment, and standards of excellence set a standard for us all."

Joshua Leeds, author of *The Power of Sound*

"Dorinne Davis is doing extraordinary work in the field of sound healing. In particular, she has studied, researched, and incorporated in her practice the work of some of the outstanding pioneers in this field. She is a wealth of information and knowledge as well as being a gifted practitioner of many different modalities that utilize sound for health and wellness."

Jonathan Goldman, Director of the Sound Healers Association, author of *The Divine Name: The Sound that Can Change the World.*

The Cycle of
SOUND

A Missing
Energetic Link

Dorinne S. Davis

NEW PATHWAYS PRESS – NEWTON, NEW JERSEY, USA

New Pathways Press
55 Mary Jones Road
Newton, N.J. 07860
973.222.3277
pub@newpathwayspress.com

New Pathways Press is an imprint of New Dimensions Publishing, LLC.

Publisher's Cataloging in Publication
Davis, Dorinne S.

The cycle of sound – a missing energetic link/Dorinne S. Davis – 1st. ed.

    p.   cm.

Library of Congress Control Number: 2012940327
    ISBN-13: 978-0-9824187-1-0
    1. Health.   2. Alternative Therapies.   I. Title

First Edition – July 2012

Editing by Sarah Collins
Cover and interior design by Deborah Perdue
Indexing by Mary Harper
Cover Photo – 2010 © Victoria Kalinina. Image from BigStockPhoto.com

Dedication

*T*his book is dedicated to the loving memory of my mother, E. Doris Taylor, who always encouraged me to think outside of the box, allowed me to explore all of my options, encouraged me to be creative, made me realize the true potential in all people, had me recognize that all people were special and had good qualities, showed me the power of alternative approaches to wellness, encouraged me to stay with a task until finished, and encouraged me to never accept closed doors. Expect the good and all things are possible.

And, as I say within this book, your name reflects an energy that propels you towards a destiny. My mother gave me a unique name because she knew I would have a unique destiny. Thank you, mom.

~ D I S C L A I M E R S ~

This book is intended to be a source of information and education. It is meant to raise one's awareness about alternative therapies available outside the scope of modern medicine. Every effort has been made to provide clear and concise information; however, being fallible, mistakes may occur in content or typesetting.

The book is sold with the understanding that the publisher and author are not engaged in rendering medical, psychological, or spiritual counseling or other professional services. If assistance is required, please seek the services of a competent professional.

The publisher and author shall in no way assume responsibility, be liable in any manner to any person or entity with respect to any loss, damage caused, or alleged to have been caused, directly or indirectly, by the information contained within this book.

If you do not choose to be bound by the above disclaimers, please return this book to the publisher for a full refund.

Table of Contents

PREFACE

CHAPTER 1: SHARING THE PAST, PRESENT, FUTURE......1
My journey into the world of sound 5
Introduction of body sound energy 13

CHAPTER 2: OLD CONCEPTS/NEW IDEAS................23
Sound-based therapy 23
The Tree of Sound Enhancement Therapy 24
The seed: Basil body rhythms..........................24
The root system: The sense of hearing.................26
Trunk: Sound processing skills 26
Lower leaves and branches: Auditory processing skills . 27
Upper leaves and branches: Academic skills 28
The head: Overall maintenance and support..........28
The Diagnostic Evaluation for Therapy Protocol 29
The concept behind The Davis Model of
    Sound Intervention 31
The founder of sound-based therapies 33
The five laws of the Voice-Ear-Brain Connection.......34
What's so special about the ear? 36
Otoacoustic Emissions 38
Ototoning...42
Silence 45
Movement and the ear...................................45
Old concepts and new ideas.............................51

CHAPTER 3: SYSTEMS AND BASIC SOUND 55
Systems and vibrational patterns of the body..........56
Whole body subtle system 62
Sound intervention.....................................72
Sound properties.......................................76
Why only the voice, ear, and brain? 79

CHAPTER 4: SOUND AND THE BODY 83
What is sound? 85
The effects of sound on the body 86
Rhythm, the seed, and energy movement 86
Natural energy phenomenon............................91
Cellular checks and balances.........................94

Cell communication. 98
Cell patterns and responses 102
Sound, music, and physical issues. 104

CHAPTER 5: THE VOICE 113
The whole person response 115
Connecting to our inner voice 117
The voice and body vibration. 120
Listening beyond the words. 121
Vocalizing and vocal techniques. 125
Dr. Tomatis and the voice. 126
Techniques that include concepts of the
    Voice-Ear-Brain Connection. 128
Spin off methods 132
The science of Human BioAcoustics 133
The amazing voice 136

CHAPTER 6: THE VOICE-EAR-BRAIN CONNECTION . . . 141
The system. 142
The voice 143
The ear 144
The brain 147
The body. 149
The foundation behind the
    Voice-Ear-Brain Connection. 151
Evolution of a new process 153

CHAPTER 7: OTOTONING 159
The process of Ototoning 173
Steps for Ototoning. 175
Steps and explanations. 176
Conclusion 186

CHAPTER 8: NEW PARADIGM. 189

ABOUT THE AUTHOR 195
SOURCE NOTES 197
INDEX. 205

Preface

*T*he Cycle of Sound: A Missing Energetic Link is the culmination to date of my life's work. It is not my intention in this book to discuss the more technical details that many people are searching for, such as: the ear, hearing, auditory perception, and the working of the brain. These pieces of information are included in my book, *Sound Bodies through Sound Therapy*. In The *Cycle of Sound*, I present hypotheses based upon actual body responses and hope for the day when we can measure the outcomes.

For hundreds and in some cases thousands of years, noted theologians, physicians, mystics, and authors have been describing and sharing the healing benefits of sound, sounding techniques, and music on the body. In many cases, each individual feels they have found the answer of how to help heal the body and how to make the person whole. I will share quotes from many of these respected persons of renown throughout the book. The work of these knowledgeable and intuitive people has provided the very core of my own work; their observations lay the foundation for the body of material this book explores. I have built upon their collection of healing knowledge by introducing the connection between the voice, ear, and brain – a *vibratory connection* – which is the missing energetic link that until now was unknown.

This book examines a new system for discovering how to obtain and remain in balance by using the *specific sounds* the body asks for. The ideas presented bring together information from many different fields: physics, biology, chemistry, sound, music, and the like. By being open to the possibilities of integrating these concepts into your life, you will discover a new approach to health and healing. It is my hope that this book provides much discussion and stimulates ideas for further research.

Dorinne Davis
January 2012

CHAPTER 1

Sharing the Past, Present, and Future

Silence is golden. What does that phrase mean to you? Maybe you remember the song, *Silence is Golden* by the Tremeloes; it was released in 1967. The lyrics suggest keeping silent about a cheating boyfriend. I can still hear the song in my head, "Silence is golden, golden, but my eyes still see. Silence is golden, golden, but my eyes still see."[1] The words consider the importance of keeping silent, but also how the eyes still notice what is happening. The brain is putting together a linguistic response while the body is responding to sensory information.

Typically, 'silence is golden' is considered a proverb, expressing the virtues of keeping silent. The origin is thought to have been first reported in English in 1831 by Thomas Carlyle, translating from the German *Sartor Resartus* (and probably the reason for the typo in the word 'silver' at the end of the quote). The character is sharing the virtues of silence:

> "Silence is the element in which great things fashion themselves together; that at length they may emerge, full-formed and majestic, into the daylight of Life, which they are thenceforth to rule...do thou thyself but hold thy tongue for one day: on the morrow, how much clearer are thy purposes and duties...when intrusive noises were shut out! Speech is too often not...the art of concealing Thought; but of quite stifling and suspending Thought...Speech too is great, but not the greatest. As the Swiss Inscription says: ...Speech is silvern, Silence is golden."[2]

The quote highlights the fact that words can be powerful but silence provides the time for things to fall into place for real understanding. According to The New Dictionary of Cultural Literacy, the phrase 'silence is golden' means, "silence is of great value".[3]   By the time you finish reading this book, I hope the phrase will hold great value for you.

Our world has become so noisy today that we have to listen carefully for the silence, sometimes actively searching for it.   Still, even in silence we often have extraneous noises.   Many of the noises we hear are electrical noises or other vibrations we may subtly hear.   We are constantly being bombarded by unknown frequencies.   Perhaps one of the reasons that the Amish population in the United States has no autistic children is because they are not bombarded by the energy of the unheard electrical frequencies.

Within our thought processes, we pause for silence.   Our thoughts are not run on words or sentences; we pause between words and sentences.   Although the patterning of our words and thoughts are important, it is also the silence that helps put our thoughts together.   Deepak Chopra writes about the deep silence in everyone, which is the origin of our intellect.   And in order for us to reach our maximum, he says we must master this inner silence and its secret places.[4]   We need silence to balance out the extraneous noise around us, and we need silence to fulfill us and hear what our body needs.   Silence also provides the space to hear the sound to help promote self-healing.

In the book Sounding the Soul: the Art of Listening, author Mary Lynn Kittelson shares the following:

> "Silence is not the same as soundlessness…   Silence is a
> meaningful experience around sound…   Silence… can have
> a weight, a pressure, that pushes emotions and contents into
> 'deep water'… (People) avoid silence, holding off what is
> happening at a deeper level."[5]

Silence can bring a person into an uneasy place where they are afraid of what they might reveal about their inner self.   Kittelson also shares that in silence, the essence of the soul is unspoken.[6]

John Beaulieu in his book *Music and Sound in the Healing Arts* reveals that the center of all sound is silence, and the deepest level of listening is within silence.

> "All sounds rise from and lead back to silence. Listening is the art of discovering silence. Silence is the key to the many adventures the world of sound has to offer. Through silence we are truly safe and free. We know the beginning and we know the end." [7]

American composer John Cage's *4 Minutes and 33 Seconds* instructs the musicians to rest their instruments during the entire piece. The piece is totally silent. Mr. Cage said that while there was no deliberate musical sound produced, there would be unpredictable and unintentional noise wherever the piece was played, which therefore created the music of the piece.[8] What a fascinating idea, especially for 1952! He demonstrated that sound and silence function well together.

Within total silence we can experience our own sound – our divine sound. All other sound experiences simply help us bridge the gaps until we get to our own inner sound, or our soul. In Surat Shabd Yoga, practitioners listen to and meditate on the inner sounds found in the silence around them. By doing so, they are able to travel to different levels of consciousness. I suggest that they could also bring themselves to a special experience by Ototoning[SM]* the sound. (This concept will be introduced in a later chapter.)

There is more to sound than just music or words. It is important to break away from the noise around and within us in order to experience the vibration of our soul.

Unfortunately, in today's technological world we seem to have regressed to where we need constant sound in our environment. When there is quiet, people often feel the need to make noise. When the sound is turned off, some people feel empty or hollow inside. They feel more whole when there is sound or music around them. Mary Lynn Kittelson in her book *Sounding the Soul: The Art of Listening* shares the following:

---

* Ototoning is a service mark of Dorinne S. Davis.

"Experiencing soundlessness – or relative soundlessness – is a disturbing experience for most hearing people, producing feelings of emptiness and deadness. They report feeling 'unnerved,' that their voices 'go nowhere,' and that they can hear their hearts beat... Soundlessness shuts us away from our acoustic background, away from the surrounding vibratory field that hearing people are so used to... Sound is a necessary 'container,' a vibrating and animating energy all around... Sound provides a sense of well-being... The presence of sound, even if distorted, offers a sense of surroundings, contact, and containment. So we need sound, we seek some sound, any sound, rather than endure soundlessness. Its effects upon mind and body are vital. Nonetheless, its importance does not mean that we are particularly conscious of sounds as we hear them."[9]

So sound can have an unconscious response within us. When the sound creates a response at the conscious level, we see it as something in which we actively participate. Then it becomes a listening activity. We listen to the sounds around us. We listen to what someone is saying. We participate in listening to nature or communicating with someone.

Tuning into the sounds within and outside our bodies is not always a conscious decision. We only consciously tune in when we need to listen – either in a protective or social mode. Developing our capacity for tuning into the silence of sound will enable us to listen to what our bodies are telling us we need. It is only through the silence that we can hear our bodies speak.

You might be wondering why all this information about silence. After all, silence is the absence of sound. Why should this be included in a book about sound? Silence is an important factor in what you will learn is the Voice-Ear-Brain Connection[SM],* and it can't be fully understood without the inclusion of silence.

---

* Voice-Ear-Brain Connection is a service mark of Dorinne S. Davis.

My Journey into the World of Sound

By profession I trained as an audiologist specializing in educational and rehabilitative audiology, working for many years in the school system and also a hospital. Along the way, I credentialed in many other educational specialties, and have taught all ages from Kindergarten to college, including supervising student teaching candidates in speech and hearing. When I was consulting at a school for autistic children I was asked to learn my first sound-based therapy, Berard Auditory Integration Training; subsequently, I have been credentialed in twenty sound-based therapies.

Friends and colleagues ask me how I became interested in sound. I have always found the healing aspects of sound to be of interest. For example, my favorite movie of all time is the original version of *The Day the Earth Stood Still*. I didn't realize until recently that the spaceman was healed with sound after being shot by the Army. I was actually disappointed that the remake of the movie did not use sound as a healing technique. Instead they used sound as a weapon.

When I was in grammar school, my mother became the school nurse. (This was before school nursing had a specific degree, training, and certification.) When it came time to administer hearing tests, she pulled out the audiometer and said to herself, "I wonder how this works?" I spoke up and said, "I bet I can show you how." Apparently I had watched the previous nurse and knew exactly what to do. My interest in audiology must have been developing! As I grew a little older, my grandfather began to wear hearing aids. I was very interested in making sure he could always hear what I was saying.

Most summers my parents would send me to summer camp. One autumn day at school, I ran into one of my camp counselors at the nurse's office. Although it was never discussed at camp, she was a teacher for hard of hearing children and she was there to help a girl in my class. I was very interested in what she was doing. Many years later while in college – I majored in Speech & Hearing as well as Speech & Drama – I ran into another camp counselor of mine. She was an audiologist collaborating on a project at the college. I ended up staying in contact with her for many years. I find it fascinating that my life was so drawn to the field of hearing and that I had so many connections with people in similar fields throughout my life. At that time, I never realized the difference between the world of hearing and the world of sound. And I never considered that the energies of the world could bring balance to our lives.

Growing up, my parents were very supportive and allowed me to explore my many interests. I always enjoyed singing in the church choir as a child and also as an adult. I love musicals and would often be cast in parts ranging from the chorus to the lead. For two summers, I worked towards my actor's equity card in a summer stock theater, which only offered musicals. I always felt better when I sang.

Additionally, I started playing instruments when I was in third grade. I can't say I ever became accomplished in one instrument because I didn't practice as much as I should have. But I did learn to play the piano, clarinet, flute, violin, and oboe. The oboe was the one I was best at, and I had a chance to study with a member of John Phillips Sousa's band. If I had known back then what I know today in regard to connecting one on one with an instrument, I would probably have stayed with music longer.

My early professional work was in a school system, first as a teacher of the hard of hearing and then as an educational audiologist. This was a new term and I am very thankful that my director at that time was open to realizing why the position was important. Later on, I became one of the founding members of the Educational Audiology Association and eventually its president.

This newly titled field of audiology allowed me to better support my students by using their hearing skills. What you do with your hearing was always much more interesting to me than the diagnostic pieces. Implementing an auditory therapy program within the school system, I often provided therapy for students who not only had hearing loss but auditory processing issues. Soon I began to notice a significant number of students with auditory processing issues who had histories of middle ear infections. My early work with these students helped formulate my ideas expressed in my first two books, *Otitis Media: Coping with the Effects in the Classroom* and *A Parent's*

Guide to Middle Ear Infections. Making sure they were able to maximize how they received and utilized incoming auditory information, I would work with these students sometimes four times a week. The change in their skills was often apparent after two years of doing the therapy that I designed. (Typically, I can make similar types of change in three months with my current approach.)

My brother introduced me to an assistive listening device that his company was producing, and after achieving great success with my students I began searching for other devices to help enhance the listening process. Ultimately, I started a company called Hear You Are, Inc., a catalog company offering a large assortment of devices to enhance listening. Using these devices with both my students in the school and the clients I saw in my rehabilitative audiology work at a local hospital, I helped many students and clients enhance their response to sound.

In 1992, I was consulting at a school for autistic children when I was asked to attend a course on a new program being introduced into the United States. It was called Auditory Integration Training and was being taught by the developer, Dr. Guy Berard. Researching the concept, I found that not too many people supported it. After taking the class I came away with many questions because there was no hard research to support what was happening or how it worked. I decided that I needed more information. It turned out that this training changed my professional life!

Years later, I realized that one of the parents of a child I worked very closely with in the school system had tried to get me interested in this very same program. Once again the energies in my life were directing me to what I am doing today. Talk about synchronicity!

In my private practice, the positive changes impacting my clients through the program propelled me to continue learning more about sound-based therapies. (I was not allowed to use the program in the school system.) I have now been trained in twenty of these therapies, many of which I wrote about in my book, *Sound Bodies through Sound Therapy.* In 1998 I closed Hear You Are, Inc. and opened The Davis Center. Since 1992, so much change was occurring for my clients with the sound-based therapies that I no longer needed the assistive listening devices for clients. No longer working with hard of hearing students, my focus became the challenges associated with auditory processing. The sound-based therapies seemed to be making change with these issues. I was transitioning from the hearing and hearing impaired world into the sound world.

As I became credentialed in more sound-based therapies, I began to see that all sound-based therapies are not on equal footing. Although each of the many different sound-based therapies that are available make change for the listener, an order for the correct administration of these many different therapies became extremely important. So, I designed The Tree of Sound Enhancement Therapy®* which I wrote about in my book *Sound Bodies through Sound Therapy*. The Tree analogy as written about in that book lists four areas of the Tree. Since that time, I credentialed in more therapies and learned that two new areas of the Tree had to be established. I will list these areas in chapter two.

When I finished writing *Sound Bodies through Sound Therapy*, it became apparent that there needed to be a way to identify the necessary sound-based therapy(s) and the order in which they should be administered. This is how the Diagnostic Evaluation for Therapy Protocol (DETP®)* developed. This test battery determines if, when, how long, and in what order any or all of the many different sound-based therapies can be appropriately administered. It is key to having maximum success with the many different therapies. The problem with using this assessment is the limitations in who can administer the battery of tests. Because it includes hearing related testing, an audiologist must administer the testing. So currently, I am the only one administering the test. People now travel from around the world to receive the test or else I travel to their area to administer it. The goal for the future is to put it into a simpler format and share it with others.

When I first opened The Davis Center, I emphasized the hearing, listening, and auditory processing pieces of what I was doing, as that

---

* The Tree of Sound Enhancement Therapy is a registered trademark of Dorinne S. Davis.
* DETP is a registered trademark of The Davis Center.

was what I knew at the time. When my unpublished research about the acoustic reflex muscle making physiological change after receiving Auditory Integration Training was complete, I made a quick review of central auditory processing test scores pre- and post-Auditory Integration Training, demonstrating growth in these scores after six months. As I was measuring the results with known audiological tests and seeing change, I shared in *Sound Bodies through Sound Therapy* what I found about the hearing and auditory processing components of using Auditory Integration Training and subsequent sound-based therapies. This was important for laying the groundwork in understanding how the different sound-based therapies make change. The therapies were evaluated by me from a perspective of seeing how the therapy impacts the auditory system. The Tree analogy had not yet been defined.

In 1999, I was trained in the newly developing science BioAcoustics™* that uses vocal analysis to identify irregular energy patterns within the body. Using low frequency sound to make corrective changes in those patterns, the body is supported towards a more natural form and function. This science provided a final piece for me in understanding how sound in general makes change for each person. And with this piece, the Tree was designed with its first model. The importance of how this science supports the body and how it helped shape the Voice-Ear-Brain Connection will be shared in a later chapter.

By using the concepts of BioAcoustics and the knowledge of the other sound-based therapies, especially the work of Dr. Alfred Tomatis, a connection between the voice, the ear, and the brain was further defined through my research that is now known as The Davis Addendum to the Tomatis Effect. This was introduced at the Acoustical Society of America in 2004 and demonstrates the importance of using not just the frequencies of the voice, but also the frequencies of the ear. This was the beginning of recognizing that there is a cycle of sound, as I saw a connection between the voice and the ear regarding imbalanced frequencies. So now, I was not only using my audiology background, but my speech pathology background as well, because I spoke of *hearing* and *the voice*.

---

* BioAcoustics is a trademark of Sharry Edwards.

They say a pioneer is one whose work is rejected by others. I remember submitting my work on spontaneous otoacoustic emissions within The Davis Addendum to the Tomatis Effect to a web portal for papers on otoacoustic emissions. I was rejected! At the present time, I am also not allowed to present my work to conferences of the American Speech Language Hearing Association (ASHA) or the American Academy of Audiology (AAA), because what I do is in direct contrast to their strongly held positions regarding speech and hearing. I remember having been scheduled to present an all day workshop where attendees could get ASHA CEU's (continuing education units), and at the last minute the conference organizers said that I had to change my topic because ASHA would not give the CEU's, and people had paid to attend already. I also have given up submitting papers to their journals because of the continuous rejection. So, I guess I am a pioneer!

The therapies I was learning had to fit into the category of a sound-based therapy. They had to use the power of sound vibration with special equipment, specific programs, modified music, and/or specific tones or beats, the need for which is identified with appropriate testing. Initially, when I needed to discuss this with other professionals I tended to revert to my knowledge of hearing, auditory processing, speech, and learning/development. I knew I was introducing sound vibration, which required the involvement of vibrational sound stimulation within the programs, but did not emphasize the importance of sound energy and its effects on the body.

Once the Tree was developed and I delved more fully into the connections between the voice, the ear, and the brain, sound energy vibration took on more importance to me. My understanding of what I had been doing evolved from my audiology and educational background, but my venture into the sound world became even more important to the development of the Voice-Ear-Brain Connection. My background only provided knowledge from which I would be able to grow, develop, and expand into the sound energy world.

It soon became important to stress the fact that there is a difference between hearing and our response to sound. Frequently, clients who have been with me for years as well as new clients, still think of the work

The Davis Center does only in terms of hearing function. But this changes once they begin to understand what I am saying and experience the changes clearly evidenced by the process.

As this book will share, hearing and our response to sound are not necessarily synonymous. Once while working as an Educational Audiologist, the high school physics teacher asked me to teach the segment on sound. I took on this task energetically and pulled together an entire week of lessons. I began with the well-known statement, "If a tree falls in the forest, and no one is around to hear it, is there a sound?" From the traditional approach of physics the answer was no, because you needed an ear to hear or register the sound. However, we now know that indeed a sound is made from any movement because the energy of the molecules will make that sound. The sound may be outside of the ear's register, but the brain would receive the vibration.

What is now known as The Davis Model of Sound Intervention^{SM}* incorporates hearing, but it is mainly our response to sound that is important for the success of any sound-based therapy. So when the many practitioners or particular therapies stress hearing, listening, vestibular function, health challenges, or other skills, the reality is that the processes changed are due to an outcome from our body's response to sound, and not the skills or challenges addressed. (This concept will be discussed further in chapter two.) The results of this model provide an approach to supporting change for the whole person. It incorporates The Cycle of Sound®,* which functions only within the Voice-Ear-Brain Connection. Because The Davis Model of Sound Intervention is used in support of the whole person, it is not appropriate for clients to say, "I only want to do Auditory Integration Training" or "I want to do the Tomatis Method®".* For a therapy to be introduced to a client, a determination must be made as demonstrated by the Diagnostic Evaluation for Therapy Protocol (DETP).

---

* The Davis Model of Sound Intervention is a service mark of Dorinne S. Davis.
* The Cycle of Sound is a registered trademark of Dorinne S. Davis.
* Tomatis Method is a registered trademark of Tomatis Development, SA.

This book will minimally provide the reader with information about the sound-based therapies. More importantly, I will share what, how, and why the body's response to sound is more important to consider than the individual sound-based therapy by itself.

Everything in the world is made up of vibration, and vibration has a sound! So everything in the world has its own unique sound. We will discuss this in great depth throughout the book. Once the concept is understood, the way you view the world around you will change. If sound is everywhere, how does our body respond? Our bodies respond to these sound vibrational energies from within and without. This can be seen as a cycle.

Once we see how the cycle exists, a brand new term, *Ototoning* will be introduced. Ototoning is an exciting new concept that takes the idea of toning – a practice that has been used for hundreds of years – to a new level. Ototoning utilizes the sounds from within the body to support making a change for that person individually. Group toning or soundings only advance the level of change to the group level; Ototoning makes change at the individual level. This concept will be explored in detail in chapter seven. With the introduction of Ototoning, my work now shifts into sound healing from sound-based therapy. This is a major advancement for the field of sound healing.

Ototoning incorporates the concepts of the Voice-Ear-Brain Connection, and provides the foundation for wellness centers using sound as the main offering. In the future, there will be certified practitioners who will offer a quick test battery to determine the necessary series of sound-based therapies that will support learning, development, and wellness changes. The therapies will be individualized and the person will be able to listen to their programs at home or in the office. The outcomes will be more directed to each person's core levels. Instead of chanting, toning, singing, and listening with others, the individual will experience what his own body needs, not what the next person needs.

As my process evolved, I was able to offer my diagnostic testing in an outreach location and the sound-based therapies as home programs. I was recently at one of the outreach locations and a local nurse practitioner invited me to dinner. She wanted to hear more about my ideas and research, and how my process worked. I shared the basic ideas and concepts with her. She listened quietly, only asking a few questions for clarification. When I finished she looked at me and paused, then said, "Wow! What you are doing goes directly to the soul level of the body!" I simply smiled at her, recognizing that she understood everything I had said. Personally, I had never put it into those words but she was right because I work with the energy of sound and get to the core of the person's issues. I get to the inner most level of the body. Edgar Cayce said it best, "each individual is ultimately a... soul, having a physical experience."[10]

Introduction of Body Sound Energy

The twentieth century viewed medicine from a Newtonian model which sees the world and body as machines comprised of many parts all working together. As the century wore on and quantum physics took shape, this model began to be integrated with another model for healing the body. This was based on Einstein's work that sees not just the body but also the entire world as complex fields of energy interacting with each other. The terms vibrational medicine and energy medicine have been applied because matter as energy ($E=mc^2$) provides the foundation for considering the body as a dynamic energy system. By manipulating this energy system the fields are rebalanced, thereby restoring order to cellular physiology. Today, this model for healing the body is not recognized as a viable medical option, yet more physicians are acknowledging the impact of subtle energy systems that can help a person heal and feel more balanced in their life. Perhaps the new way of looking at health and wellness is to look at the body in balance as opposed to the body in a diseased state.

Scientists have been studying for many years how mammals use sound to communicate and obtain information about their environments. Dr. Arthur Popper from the University of Maryland suggests that the auditory or 3D effect of sound surrounding a marine mammal provides important information for the mammal.[11] Within this 3D area, the mammal can detect sound important for critical survival and communication signals. And anything within this area that alters the mammal's ability to detect these signals poses a possible threat to the mammal's life and survival.

It is my assumption that humans respond the same way. We have an energy field or 3D area surrounding us. When a disturbance in the energy occurs, the disturbance is felt by our bodies creating irregular wave patterns disseminated throughout the body. Even though the disturbance could be life threatening, it might also be so minimal that it is not brought into our conscious awareness, yet it has a significant impact on us such as with the onset of an illness or disease.

The terms vibrational and energy medicine are used interchangeably, but for consistency I will defer to vibrational medicine since I work with sound which people tend to think of in waves, frequencies, or vibrations even though sound is also an energy source. Vibrational medicine then attempts to make change for the person by changing their energy. Our cells and molecules are constantly moving, creating energy. Our bodies are complex networks of energy fields. The life force comes from an organized network of subtle energy systems. These systems are established from interwoven energy fields that exist at the cellular level. Our circulatory system, blood system, and hormonal system are some of the more recognized systems. But there are more in-depth systems that also coordinate electrophysiological connections and the cell structures themselves. The more in-depth or subtle the system is, the deeper the connection with the person's soul. The soul dimension of the person is the energetic framework for the functioning of their physical being. The connection between soul and physical body is understood in the relationship between the body's subtle energy systems and the matter of the body (the physical body).

Many of my predecessors who discuss the person's soul often link spirit and soul as coming from this inner core. For the purposes of my work, I consider the soul to be within the inner energetic framework and the spirit to be stretching outward beyond the physical body reaching out to the universe. The spirit vibrates outward from the inner energetic framework, thereby linking soul and spirit. Throughout this book, you'll recognize that I believe we have to balance the soul first and then balance the spirit in order to balance those around us as well as the universe. To me, our soul is personal and our spirit is something we share. To balance both creates universal harmony.

In his book *Vibrational Medicine,* Dr. Richard Gerber says:

"Vibrational medicine is a systems approach based upon the Einsteinian paradigm of healing. Vibrational medicine attempts to interface with primary subtle energetic fields that underlie and contribute to the functional expression of the physical body." [12]

The connection between the voice, the ear, and the brain is a major subtle energy system that has not been recognized or discussed in its entirety until now. This book will discuss not only the connection between the voice and the ear, but will introduce the energy system that exists between the voice, the ear, and the brain.

The subtle energy systems of the body seem to be composed of matter with different frequency characteristics than that of the physical body. Within physics there exists the Principle of Nondestructive Coexistence that says the energy of different frequencies can coexist within the same space without destroying each other.[13] This means that the energy within our subtle energy systems can work in tandem with our physical body. The subtle energy system of the Voice-Ear-Brain Connection can exist and support the working of the physical body in much the same way that the circulatory system supports the physical body. The subtle energy systems synergistically combine to support the functioning of the physical body.

If the circulatory system isn't working efficiently or correctly the physician works to improve this system. If the skeletal structure develops problems then chiropractors, physical therapists, and/or specialized physicians all work to effect change. And if the system connecting the voice, ear, and brain is not working, the specialized sound therapist works to improve this system. It is important to make change at the subtle energy level in order for the change to be effective at the physical level.

Within the subtle energy system levels, some say the subconscious exists. Some call it the unconscious, but the body is to some degree conscious of what is going on at that level so I will call it the subconscious. Edgar Cayce shared with many that there was an "inextricable connection between vibration and consciousness".[14] There is often an emotional connection established at this level based upon the learned responses of our cell energy early in life or from past lives.* These learned emotions often allow our body's responses to influence not only our personalities but physical disease as well.

In Edinburgh, Scotland the European Science Foundation held a symposium in 2008 on understanding the mystery of human consciousness. In closing comments, scientist Ezequiel Morsella suggested, "consciousness is the result of competing demands on our skeletal muscle system, and is thus an advanced form of mediation between the various systems which constantly seek to control the body".[15] This radical proposal of the time suggests a cycle of systems working together within the body trying to find a balanced position. Where and how does the conscious aspect of a human being begin and the subconscious release what is happening within? At what level of our internal vibrational systems does the conscious response become apparent? At what level does the subconscious release its hold and allow the conscious level to take hold of an idea or response? Does the vibration at the subconscious level influence our conscious response? Does the cyclical interaction of our systems influence this differentiation? At what level does body control occur? How and where are these responses demonstrated in our human reactions or overall wellness?

In *Vibrational Medicine,* Dr. Gerber discusses an etheric body which he describes as an energy interference pattern that exists within and without the body.[16] This description allows us to imagine energy being pulled into our deepest recesses and stretching out beyond the body as far as allowed. This etheric body represents the sound energy system that will be described later in the following chapters.

Although beginning at the subconscious level, the system within the connection between the voice, ear, and brain seems to bridge the gap between the physical body and the etheric world. By reconnecting or

---

* For more information about how our cells learn emotions, behaviors, and everyday responses, refer to Dr. Bruce Lipton's work, *The Wisdom of Your Cells.*

repatterning the energy between the voice, ear, and brain, the physical body is brought to a better natural starting place to learn, develop, and maintain one's health and wellness. The give and take of the body's cell energies are vibrationally at a better centering place. The repatterning is accomplished with the use of specific sound-based therapies and can be supported or complemented with less specificity with sound healing techniques. With these therapies, emotional blockages have been released and/or emotional connections have been enhanced. The individual is more aware of the important subtle patterns of sound necessary for positive interaction with the world – patterns of rhythm, duration, intensity, and frequency. The world they live in becomes more alive.

Sound is the energy of the universe. Every atom, molecule, and sub-particle vibrates. Any vibration creates a sound, as does a frequency. Everything in the universe has a frequency and vibrates, and therefore emits sound. At UCLA, a researcher discovered that every cell in our body emits a sound.[17] Researchers have also discovered that inanimate objects vibrate and emit frequency, therefore sound.[18]

Investigators at the University of Copenhagen are beginning to question whether the nervous system actually operates from electrical stimulation, and hypothesize that perhaps it really works with sound stimulation.[19]

Many scientists and healers have discussed how the universe was created by sound. Theologians reference the Gospel according to John in the New Testament as saying, "In the beginning was the word..." The 'word' has been discussed as the initial sonic vibration of the world. If everything has a sound, then this initial sonic vibration began the process.

With human evolution, language (as the sound) was formulated. Prior to languages, humans still vocalized. We vocalize in many ways: humming and chanting for instance. Over time, generations have learned the power of using mantras or toning techniques to support a connection between the mind, body, and spirit. All of these techniques have found positive responses for people who are able to get past their subconscious imbalanced reactions.

In some forms of yoga, the highest level of atonement is to reach the soundless sound level, where the sound can be made and heard in the brain without physically making the sound verbally. Except for this highest level, how has the body vocalized? With the voice!

Children have fun playing with their voice and sound; they learn language from getting feedback from their vocal play. Cave dwellers made grunts and other sounds to express themselves. When we are hurt, we cry out in pain. When we feel good, we often hum or sing a tune. We release our emotions and physical needs through our voice. As every frequency produced by the voice is associated with a specific component in the body such as a muscle or biochemical, we now can identify the body's imbalances through vocal analysis.* Our voice is reflective of who we are and also provides a way for us to repattern our imbalanced frequencies.

Instead of simply toning a generalized sound as people have done for centuries, Ototoning helps the person develop their own sound to correct their immediate imbalance. Eventually, with the expansion of this new concept, Ototoning will also let the person know what their specific frequency issues are so that they can utilize the appropriate frequency to self-heal or visit a physician. The process is one of wellness balance. What may result in a total healing for some may only relieve certain properties of the imbalances for others. Although the process may relieve stress or disease for many, the concept of Ototoning cannot be considered a cure for any issue until further research is conducted.

There is reciprocity within any sound cycle. We have the give and take of energy, and we see polarities of energy expression when we talk of matter and spirit. While this reciprocity demonstrates the opposite's effect, it also demonstrates how the energy complements each other. Everything has an expression of frequency and a reception of frequency. So the body gives out a sound and takes in a sound at every cellular, every atomic, and every subatomic level. Many different fields discuss some form of receptor or expression points. For example, biologists discuss cellular cycles while physicists discuss various cycles within the laws of nature. Energy techniques incorporate

---

* For more information review the research of Sharry Edwards and the science of Human BioAcoustics.

the cycles of the body. Everyone is discussing the same thing. Let's unify these pieces by discussing the expression and reception of sound frequency.

Everybody knows that our ears are our hearing mechanism. But the ear is actually much more than that. The ear also emits a sound called an otoacoustic emission. This concept was first identified by Dr. Gold in 1948 and reintroduced by Dr. Kemp in 1978. Their work has resulted in diagnostic testing that helps identify many auditory dysfunctions. Our body's spontaneous otoacoustic emission is still being researched. Many are unsure why we have such a process in our body. However, because of the emission and reception processes for everything in the body, understanding why we have spontaneous otoacoustic emissions in addition to having hearing receptor cells begins to take on new meaning to our body's cycle of sound. Perhaps these spontaneous otoacoustic emissions provide greater information than hearing function. What other information can these emissions provide?

Later on, we will re-examine The Davis Addendum to the Tomatis Effect, which connects the imbalanced frequencies emitted from the ear with the imbalanced frequencies emitted with the voice, and supports how the body wants to self-heal with vocalization through Ototoning. The voice regains coherence with the introduction of the correct vocal tone identified from the ear's emission. This introduces a new energy-field system called the Voice-Ear-Brain Connection. By establishing coherent energy patterns with the voice through Ototoning, the sound can influence the energy events happening subconsciously at the cell or subatomic level to help the body establish a more normalized form and function. This repatterning provides a better connection between the inner soul level and the spirit level of the universe. Ototoning appears to provide this link and supports the importance of the Voice-Ear-Brain Connection for every human being. In addition to the evolution of my work with sound, this book will examine how the Voice-Ear-Brain Connection has evolved as an important system for the body in balancing our lives.

Sound Notes
Chapter One, Summary Statements:

- Silence is golden. You will pick up more about the world around you, your own energy, and what will make you feel better by tuning into the silence.

- The goal for the future is to design a simpler test battery based on the Diagnostic Evaluation for Therapy Protocol (DETP) that can be available to interested practitioners.

- The Davis Addendum to the Tomatis Effect demonstrates the importance of using not just the frequencies of the voice, but also the frequencies of the ear.

- The Davis Model of Sound Intervention incorporates three tenets:
  1) There is a connection between the voice, the ear, and the brain;
  2) Every cell receives and emits sound; and
  3) Our ear is a global sensory processor.

- The Cycle of Sound functions within the Voice-Ear-Brain Connection.

- Ototoning utilizes the sounds from within the body to support making a change for that person individually.

- Physicians are recognizing the impact of subtle energy systems that can help a person heal and feel more balanced with their life functioning.

- Our life force comes from an organized network of subtle energy systems.

- The soul dimension of the person is the energetic framework for the functioning of their physical being.

- The connection between the voice, the ear, and the brain is a major subtle energy system that has not been recognized or discussed in its entirety until now.

- Learned emotions allow our body's responses to influence our personalities and physical disease.

- The connection between the voice, ear, and brain bridges the gap between the physical body and the etheric world.

- By repatterning the energy between the voice, ear. and brain, the physical body is brought to a better natural starting place to learn, develop. and maintain one's health and wellness.

- With the use of sound-based therapies, the individual is more aware of the subtle patterns of sound necessary for positive interaction with the world: patterns of rhythm, duration, intensity, and frequency. The world that they live in becomes more alive.

- The Voice-Ear-Brain Connection can be considered a new subtle energy field system within the body.

- By establishing coherent energy patterns with the voice through Ototoning, sound can influence the energy events happening subconsciously at the cellular or subatomic levels to help the body establish a more normalized form and function.

CHAPTER 2

Old Concepts/New Ideas

*T*he Davis Model of Sound Intervention is a new idea developed from old concepts. How has this model evolved? Each step of my growing process within the sound field has produced something special that has provided meaning to the evolving field of sound-based therapy. This chapter shares the evolution.

Sound-based Therapy

What is sound-based therapy? I define this term as using sound vibration with special equipment, specific programs, modified music, and/or specific tones/beats, the need for which is identified with appropriate testing. I realized that I was using more than sound healing techniques. Using specific programs and equipment I was presenting sound in specific ways and making major changes in the person's functioning. My approaches were being utilized by other practitioners as independent therapies; however my audiology background was demonstrating to me that there was more to this field than lumping all the therapies using sound into one generic approach. Sound-based therapy goes beyond hearing, beyond auditory processing, and beyond music. It utilizes the sound vibrational energy of the person and supports change within the person's natural form and function by repatterning this energy.

Berard Auditory Integration Training, the Tomatis Method, BioAcoustics, Fast ForWord®, Interactive Metronome®, REI, and EnListen® are just some of these therapies and were discussed in detail in *Sound*

Bodies through Sound Therapy. They are not relevant to the Voice-Ear-Brain Connection except in their proper administration for maximizing how the body uses vibrational sound stimulation.

The Tree of Sound Enhancement Therapy

In 1992, I began my training in what is now known as Berard Auditory Integration Training (AIT) with Dr. Guy Berard. Although I was told of the types of changes that could occur with this method, my questioning mind wanted more answers. Initially, unpublished research began with the stapedius muscle in the middle ear, and when I repeatedly made change in the response of this muscle using AIT, I realized something special was happening. I also saw that changes were occurring with central auditory processing skills, but not significantly. These early results encouraged me to get trained in more of what I began to call sound-based therapies. I am now trained in twenty of these therapies as a result of trying to find out how they worked and what results were possible.

As I began to see that many of the therapies were different, The Tree of Sound Enhancement Therapy model was developed. I initially introduced this Tree analogy in my book, *Sound Bodies through Sound Therapy*, but the Tree at that time only had four parts. My quest for knowledge with these therapies has now led me to realize that six parts are needed. This new Tree analogy is now brought into the Voice-Ear-Brain Connection in the following manner.

The Seed: Basal Body Rhythms

Our body rhythms start at conception and develop as the fetus grows, and then mature and stabilize as the baby is born, grows, and develops. The Seed of The Tree of Sound Enhancement Therapy relates to our core or basal body rhythms. These rhythms refer to our heart rate, breath stream, and cellular patterns. We all inherently have these rhythms directing our every day functioning. Occasionally, these rhythms do not support our ability to function at our greatest level in learning or developmental tasks, or help us maintain our wellness. Assistance with therapy at this level of the Tree can be addressed at any time before or

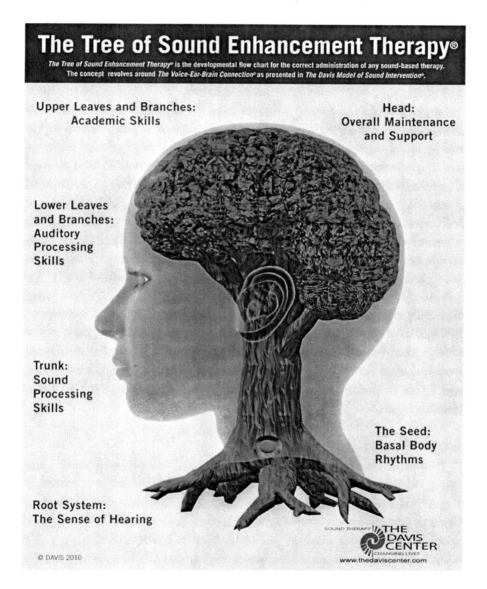

The Tree of Sound Enhancement Therapy®

The Tree of Sound Enhancement Therapy® is the developmental flow chart for the correct administration of any sound-based therapy. The concept revolves around *The Voice-Ear-Brain Connection®* as presented in *The Davis Model of Sound Intervention®*.

Upper Leaves and Branches:
Academic Skills

Head:
Overall Maintenance
and Support

**Lower Leaves
and Branches:**
Auditory
Processing
Skills

Trunk:
Sound
Processing
Skills

The Seed:
Basal Body
Rhythms

Root System:
The Sense of Hearing

SOUND THERAPY **THE
DAVIS
CENTER**
CHANGING LIVES
www.thedaviscenter.com

© DAVIS 2010

after use of the other sound-based therapies. Sometimes, because this level of the Tree is our foundation, the only recommended therapy is at this level as it introduces rhythmical patterns to which the body responds physically and emotionally.

The Root System: The Sense of Hearing

As a seed grows, roots are established. All seeds have roots, but how healthy are they? For people to function at a maximum level roots must be strong, well fed, and remain supportive for their lifetime. The Root System of The Tree of Sound Enhancement Therapy is connected with the development of the ear, specifically how the acoustic reflex muscle in the middle ear cavity supports reception of sound frequency in the cochlea, as well as in the vestibule and semi-circular canals of the inner ear. The sense of hearing is addressed at this level. The muscle reflex can be compromised at any time in a person's lifetime, meaning that it may be over-reactive, under-reactive, or not have enough muscle tone to support good responses to sound vibration. Some commonly seen causes for this may be early middle ear infections, weak immune systems, subjection to loud sound, or poor general muscle tone. The type of therapy used at this level of the Tree is Auditory Integration Training. When issues associated with one's sense of hearing are evidenced through testing, it has been my clinical experience that this level of therapy typically must be addressed prior to some of the other sound-based therapies for optimal benefit. These therapies are modeled after the work of Dr. Guy Berard, a respected French ear, nose, and throat physician. The changes that take place are a result of better sound reception.

Trunk: Sound-Processing Skills

When the roots are healthy, the tree begins to grow its trunk. The Trunk reflects how our bodies process sound vibrations. Sound is processed not only through our cochlea in the inner ear, but also through our bone structure, cell network, nervous system, and circulatory system. Sound processing is more than hearing; the vibrational energy of sound waves and patterns travel to and through our body. The connections between the voice, the ear, and the brain are established at this level by stimulating vestibular function, language skills, and attention/ focus/organizational needs. Special music (music that is altered in how you listen to it) and sound are processed through the air in the ear canal, and/or by the bones of our skeletal system to help repattern the body's foundational responses. Once the process has begun, the person's voice is introduced. The voice establishes a type of beneficial feedback that helps the listener stabilize and enhance the processing of all incoming sound with better clarity and understanding. The therapies used at this level are called Listening Training Programs and are modeled after the work of Dr. Alfred Tomatis, a noted French ear, nose and throat physician. These programs must include all parts of Dr. Tomatis's model, not just some parts of the model. This level of the Tree must be strong and supportive in order to maximally develop the integrated communication, academic, social, emotional, attention, and vestibular skills needed for learning and development. These learning and developmental skills can be present at any age level and at any time during a person's lifetime. The changes evidenced are a result of the vibrational stimulation through the parts of the ear: cochlea, semi-circular canals, and vestibule, as well as the bones, circulatory system, nervous system, soft-tissue network, and cell structure of the body.

Lower Leaves and Branches: Auditory Processing Skills

As the trunk is growing, the leaves and branches begin to sprout and grow. A healthy tree has strong branches with healthy leaves that demonstrate its overall beauty. The branches must be strong to withstand nature's surprises and everyday wear and tear. The leaves should not wither from lack of nutrition or support through the trunk of the tree. The skills at this Lower Leaves and Branches level of The Tree of Sound Enhancement Therapy are specific to auditory processing skills such as memory, discrimination, sequencing, and localization. These skills are inherent in how the brain receives information from the auditory pathway to the brain, specifically from the cochlea to the auditory reception centers in the brain, and are very necessary for the development of academic-based skills as well as attention and listening skills. Without them being well established, learning and listening can be frustrating and/or difficult. There are many therapies that support positive change at this level, either in a specific skill area such as temporal sequencing, or with general overall auditory processing skills.

However, they should only be introduced when the lower levels of the Tree demonstrate sufficient support and skill; otherwise, the person may develop many splinter skills, not integrated skills.

Upper Leaves and Branches: Academic Skills

As the Leaves and Branches maintain their healthiness, the final academic pieces can be processed more easily. A person can develop rudimentary academic skills, have difficulty with types of skills like phonics, and stay focused for short periods of time without having the earlier growth of The Tree of Sound Enhancement Therapy be its healthiest. To maximize the academic skills such as reading, hand-writing, spelling, and math, the flow of information from the Roots through the Lower Leaves and Branches must be strong and supportive. One can teach these basic skills with tutoring or special programs, but unless the information is flowing easily to understand simple basics such as the rhythm of language, the movement of handwriting, or the discrimination of phonemes, the skills are not fully integrated. Once the foundational skills associated with the various levels of the Tree are connected and stabilized, the academic skills develop on their own, and additional supportive service needs are minimized. If therapy is needed at this Upper Leaves and Branches level, the program must incorporate auditory, visual, and language skills within the program.

The Head: Overall Maintenance and Support

The developing tree is contained within the Head because of the importance of the connection between the voice, the ear, and the brain. The voice reflects the inner harmony of the body. Everything in the body has been identified from the frequency emissions of the cell structures to the emissions from the inner ear. These frequencies are reflective in the voice and through vocal analysis. Frequency Equivalents* associated with various wellness challenges help support the body towards a more natural form and function. This portion of The Tree of Sound

---

* For more information about a Frequency Equivalent or a frequency that is the equivalent of the vibration of certain compounds, elements, or organs in the body, review the work of Sharry Edwards.

Enhancement Therapy is labeled overall maintenance and support because the voice is the tool for change. The therapies at this level use vocal analysis to initially identify various wellness imbalances. The therapies then support change by listening to special sound protocols, or by using a technique called Ototoning that allows the person to use his own voice to support their own natural change.

The Head further demonstrates how the voice reflects what the ear hears and emits, and the brain receives, perceives, and sends to the body for its use. The overall Tree analogy demonstrates the Voice-Ear-Brain Connection, a process whereby we use sound coming into the body and sound emitted outward from the body to keep the body in harmony in a *cyclical* pattern. This process was finally identified through my research now known as The Davis Addendum to the Tomatis Effect, which will be shared later in this chapter.

The Diagnostic Evaluation for Therapy Protocol

Back in 1998, the idea of the Tree began to be integrated into my work as a way for clients to understand why I used different sound-based therapies. I wrote about this in *Sound Bodies through Sound Therapy*. Eventually, I discovered something important missing in my process: a way to figure out how to determine which therapies were needed, and in what order. This is how The Diagnostic Evaluation for Therapy Protocol (DETP) was created.

This diagnostic evaluation has been integrated into all four sections of the original Tree. Issues at the Roots are identified by the Hearing Sensitivity Test, which incorporates the foundation of a complete audiological evaluation, but is not testing for the same thing. Trunk issues are identified by the Sound Processing Test that is similar, yet different from a Tomatis Listening Test. Leaves and Branches issues are identified by a functional test of auditory processing skills. Associated Head issues are identified through vocal analysis. Because of the response level of the many people evaluated, there is a partial as well as a full version of the test battery available. The portable, modified version is offered to those clients unable to travel to the Davis Center. It is my hope that in the next few years this test battery will be in an easier format and available to many practitioners.

How is the information from the test battery utilized? We assume because we have learned to read, write, socialize, ride a bike, and sing a song among many other things, that we are whole. Each one of us is unique so we see many differences between us as humans. However, within those differences, the degree does not have to be that great if we think in terms of the Voice-Ear-Brain Connection. We should be receiving sound at an optimal level with our supportive middle ear muscles. That sound should then be received in the cochlea and sent to the brain with the best input, then on to the entire body via the nervous system, circulatory system, and soft tissue network. The transmission to the brain and the body for the perception of that sound should also be intact. So what is *whole*? Whole means that we receive, perceive, and integrate sound vibration, then express sound vibration for a continual balanced cycle of response. We assume this process is automatic. Yet one little instance in life at any time during our life span may repattern this cycle and create an imbalance. Here are some examples of how that may occur:

- A middle ear infection creates a distortion in the perception of speech sounds;
- A death in the family creates an emotional blockage;
- A medication creates a negative reaction; and
- Loud noisy situations trigger anxiety.

Each individual tries to overcome the change, even if it is negligible, by compensating. In most cases, this compensation then becomes the norm. With more and more disruptions to the cycle, the body finds it harder to be in balance, so symptoms occur.

The Diagnostic Evaluation for Therapy Protocol (DETP) evolved as a way to determine how this cycle of sound impacts the person. It follows The Tree of Sound Enhancement Therapy flow chart, but the results demonstrate where the body broke down and compensated within the automatic energy flow of our Voice-Ear-Brain Connection. The results demonstrate the dyssynchrony within the body as a result of the body trying to self-correct. Most importantly for the Voice-Ear-Brain Connection, the body has difficulty totally compensating without the help of a sound-based therapy or the use of Ototoning.

Because sound-based therapies are not readily available to most people, the compensated person begins to develop the symptoms of the disruptions to their system. Symptoms are too numerous to list because each person's symptoms are unique to him or her. If we clump them into types of symptoms, they can run the gamut from disease, learning challenges, developmental issues, physical responses, emotional imbalances… to any possibility. The list can be endless.

People tell me that they use my Tree with their clients. They start with the Roots and progress to the Trunk, and so on. They assume they understand the Tree, but there is more to it than simply advancing to different levels. I may suggest someone begin with the Seed or the Head before going to the Trunk, or I may suggest starting at the Roots then immediately go to the Head. The results of the Diagnostic Evaluation Therapy Protocol (DETP) demonstrate where the imbalances are; therefore, the interpretation of the results is most important. When the full product becomes available, the protocol of which areas of the Tree to use will be an automatic outcome of using the test battery. And the interpretation of the DETP will be built in so that there will no longer be any guesswork needed. It has taken twenty years to pull the information together so far and it has been an evolving process. Until the test battery is available, general sound therapists are cautioned on interpreting the Tree analogy simply by its graphic. The proper sound-based therapies should be determined only from appropriate testing. It is from this proper determination that the dyssynchrony or imbalances in the body's cellular flow of sound energy between the voice, ear and brain can be properly identified, and then guided towards a path of self-correction through either sound-based therapy or Ototoning.

The Concept behind The Davis Model of Sound Intervention

There are three concepts behind The Davis Model of Sound Intervention:

1) There are five laws that connect the voice, the ear, and the brain. Summarized, the voice produces what the ear hears, and the ear emits the same stressed frequencies as the voice. If one is modified, the other changes, and the brain then

sends the correcting response to the body. These laws are known as The Tomatis Effect and The Davis Addendum to the Tomatis Effect. I will introduce these more in detail shortly.

2) Every cell in the body emits and takes in sound frequencies, thereby portraying the body as one's Signature Symphony of Sound. If there is an out of tune frequency the instruments (cells) of the body are not supporting the symphony and the music is discordant. The out of tune frequencies are identified. The sound is introduced to the body and the brain then introduces the correcting frequency to the cell to tune it up. This will be discussed in chapter four.

3) The ear is viewed as the body's global sensory processor and not just a hearing mechanism as all of the body's senses are stimulated either directly or indirectly through the ear. Sound vibration is also processed through bone response, cell response, the nervous system, the circulatory system, and the soft tissue network of the body. The body as a whole is considered a vibrational frequency entity. This was discussed in great length in *Sound Bodies through Sound Therapy* and will be briefly discussed later in this chapter.

Once The Diagnostic Evaluation for Therapy Protocol is administered, the interpretation for the therapy protocol is considered not only from the Tree per se, but also by interpreting the integration of these three concepts. What do the person's symptoms, their case history, and their test results suggest for a sound-based therapy protocol?

With this model, the diagnosis of the person seeking to make change does not matter, for it is the sound energy patterns of the body that determine the possibilities for change in each individual. Sound-based therapies compel the evaluator and the client to focus on the energy patterns of the body, not the disability – neither medical nor educational. The disability or challenges that are helped by the therapies are as a direct result of the three key points within the underlying concepts of the process. The symptoms, disabilities, and challenges are not what are addressed. Instead, each person is helped as an individual energy entity.

Sound-based therapies hold great potential for creating significant learning, development, and wellness changes for any individual. The paradigm established with The Davis Model of Sound Intervention provides an opportunity for each person to make changes naturally by repatterning their energy.

The Founder of Sound-Based Therapies*

Dr. Alfred Tomatis is considered the founder of all sound-based therapies. He began his work in the 1940's and '50's and was the first to define the distinct difference between hearing and listening. He felt that the brain receives more stimuli from the ears than from any other organ.

Dr. Tomatis stressed the connection with the face. He mentioned many neurological connections between the voice, the ear, and the brain. The facial nerve innervates the muscles of the face, including the lips. These muscles are important for intelligibility of speech, and the clarity of one's voice. This same nerve also innervates the stapedius muscle in the middle ear, and the muscle that opens the mouth, the digastric muscle. The trigeminal nerve connects to the tensor tympani muscle in the middle ear, as well as to the masseter and temporal muscles that allow us to chew and close our mouths.

Furthermore, Dr. Tomatis believed that listening was not a given at birth. Hearing is a given (provided the mechanism is working). Listening builds with training. After one's physiological development one's hearing basically stays the same (with some exceptions like progressive hearing loss). However, one's listening capacity is learned, trained, and can change with time.

After working with opera singers and factory workers, Dr. Tomatis wanted to find a way to enhance listening skills. This led to the development of a more sophisticated machine called the Electronic Ear.

---

* A summary of the first part of Chapter 5 of *Sound Bodies through Sound Therapy* by Dorinne S. Davis.

While treating the ear and the voice he soon discovered that he was also treating changes in behavior, especially at the level of communication. He was interested in the psychological aspects of the listening response so he began to work within the areas of learning disabilities, integration of foreign language, communication balance, psychological balance, and the re-energizing of older people.

The Five Laws of the Voice-Ear-Brain Connection

As mentioned within the concepts of The Davis Model of Sound Intervention, there are five laws that support a connection between the voice, the ear, and the brain. Dr. Tomatis created a technique of listening that he demonstrated to the French Academy of Sciences in 1957. It became known as the Tomatis Effect. He stated three laws:[1]

1) The voice only contains the harmonics that the ear can hear;
2) If you allow the ear to hear the distorted frequencies of sound that are not clearly perceived, these are immediately and unconsciously restored into the voice; and
3) The imposed audition sufficiently maintained over time results in permanently modifying the audition and phonation.

Simply stated, these three laws relate that the voice can only produce the frequencies that the ear can hear, or the voice can only produce what the ear hears. This result had a major impact on the scientific community at the time. Dr. Tomatis was able to demonstrate that the larynx and cochlea, the two organs for voice and hearing, are part of the same neurological loop. Changes to one impact the other. If someone has normal hearing and they are not an active listener, the voice will not respond at all frequencies. One must consciously decide to listen to something or someone.*

Think about the following statement: *The voice produces what the ear hears.* If you ever hear a deaf person speak you can immediately pick up that they do not hear a frequency and therefore cannot produce that frequency. In effect, this happens to each one of us but to lesser

---

* For more detailed information on Dr. Tomatis, the Tomatis Method, and the Tomatis Effect, review Chapter 5 in *Sound Bodies through Sound Therapy* by Dorinne S. Davis.

degrees. Our voices reflect what our ear processes, and as you will see, can reflect imbalances within our wellness capabilities.

In 2004, I presented to the Acoustical Society of America my research that is now known as The Davis Addendum to the Tomatis Effect. This addendum has two laws that state:

1) The ear emits the same stressed frequencies that are emitted by the voice; and

2) When complementary or supplementary frequencies of stressed frequencies are introduced via sound vibration to the ear, vocal patterns regain coherence.

Exploring the connection between the voice, ear, and brain via a research study, I knew that the voice produces what the ear hears so I decided to look at vocal output and reception of sound at the ear, reflective of the work of Dr. Tomatis. Knowing that the ear emits a sound, I also looked at the vocal output and emission of sound at the ear, reflective of the work of Sharry Edwards and the science of Human BioAcoustics. (I presented a collaborative paper with Ms. Edwards in 2002 at the American Academy of Audiology on this connection.) By combining the work of these two respected individuals, I hypothesized that the connection between the ear and the voice would be the same whether it was in the reception or expression of sound. By finding a common denominator for the frequencies, I found this connection.

One hundred percent of all subjects tested had at least one matching connection between the ear and the voice. The results allowed me to hypothesize that the voice produces what the ear hears (Dr. Tomatis) and the ear emits the same stressed frequencies that are emitted by the voice (Davis, 2002).[2] This is the concept behind The Davis Addendum to the Tomatis Effect.

For over fifty years, people have been discussing the work of Dr. Tomatis and using his three laws as a foundation for supporting change with listening, development, emotional connections, response to sound in general, and more. The last two laws are the final pieces in the use of sound received and expressed by our body for our full, total body enhancement. These two additional laws help one understand the Tree of Sound Enhancement Therapy and most importantly the Voice-Ear-Brain Connection. The full five laws demonstrate The Cycle of Sound.

The voice as identified within these five laws is the key to understanding each individual. The voice reflects the current imbalances through vocal analysis or by simply listening to the person's voice. And the voice is the key to reintroducing the needed sound to the body, allowing each person's energetic functioning to balance itself naturally for both physical and emotional well-being. Additionally, with the technique known as Ototoning, one's voice will reintroduce the needed sound identified from the ear to the full body, working towards coherence between the voice, ear and brain.

What's So Special about the Ear?

The ear is an amazing part of the body. It is the only sensory system fully functioning in utero. By four and a half months (in utero), the ear is working. The neurological system of the ear is developed by seven months (in utero). Our sense of hearing is also the last sense we lose before we pass away. Equally amazing is the fact that the three little bones in the middle ear cavity called the ossicles are the only bones in the body that are fully matured at birth.[3] Their size does not change as they grow and develop.

The ear is made up of three basic parts: the outer, middle, and inner ear. The outer ear picks up sound and directs it down the ear canal. At the end of the canal it vibrates the eardrum, which in turn activates vibration through the three tiniest bones of the body. From the third little bone, sound then travels into the inner ear that is also made up of three parts: the cochlea, the semi-circular canals, and the vestibule. Typically these last two are thought of together and discussed as your vestibular function and motor skills. The cochlea picks up sound vibrations and sends them to the brain.

All of our senses are vibrationally impacted through the nerves that pass through the ear. Consider sound waves traveling to the ear. Think of the vibrations of sound triggering responses in waves traveling outward from the ear via our nerves. (I'm referring to sound wave movement here, not neurological stimulation.) Ten of our cranial nerves pass through the ear.* These nerves will vibrationally reach most of the body and can subtly affect all sensory sensation.

Which sense is the most important to you? Is it sight or hearing? Do you say, "I see what you are saying?", or do you say, "I heard what you said?" Is what you see what you accept? Or do you listen for the subtleties of what you have heard? The eyes are in front of us and therefore seem to be the initial sense used in perceiving our world. We live in a world of vision-needed technology: television, computers, movies, DVD players, and more. We are concerned about the appearances of our body, our homes, our cars, and our social connections. The ears however, are on the side of our heads and to some appear to be a second line of defense, so to speak. The ears though, help define our lives beyond appearance. They help process the in-depth meanings of life. They help us sense the rhythms of life that provide needed balance in our day to day lives. They let us experience the resonance of the harmonies around us: nature sounds, children laughing, loved ones talking, and so much more. The ear's sensory receptors can pick up more subtle changes than our other sensory receptors. They provide very minute, subtle, interpretive, informational pieces that we may not get from any other sense. These receptors process information from *all* sound stimulation, even the tiniest atomic movement.

There is a hard-working subtle energy system within us that provides vital information. It lets us know the emotional needs of others – that the words someone is saying do not reveal their total thoughts, even when we need to fear for our lives – simply by responding to the sounds around us. It is only when this subtle sound system needs to trigger a response within our body that our consciousness of the sound messages is activated. Therefore, it is through subtle sound stimulation that the consciousness of our actions becomes participatory. In other words, we react once the subtle sound receptors tell our conscious mind to trigger an active response. The ear and brain have communicated; the vibrational sensations through the

---

* For a detailed description of the parts of the ear and how sound is stimulated through the nervous system, read Chapters 2 and 3 in *Sound Bodies through Sound Therapy* by Dorinne S. Davis.

nervous system, circulatory system, skeletal system, cellular network, and soft tissue network have responded.

Otoacoustic Emissions

Otoacoustic emissions are sounds emitted by the ear without any external sound stimulation. They are measured as single pure tones at fixed frequencies along a graphic continuum and are indicated by their amplitude or height of the response. Ears can emit more than one spontaneous otoacoustic emission at a time. The source of their origin has not yet been determined, but the general consensus among researchers is that they are sounds produced by the outer hair cells contained within the cochlea of the inner ear. The cell's cytoplasm is pressurized and movement occurs along the cell's lateral walls. The hair cell motility is a key mechanism in the ability of active cochlear tuning. The motor response in this region surpasses any other known motor response in a human being because the response extends into our upper limit of hearing.

Why do we have otoacoustic emissions? Most researchers have not considered this question. Diagnostic testing known as Distortion Product Otoacoustic Emission Testing or Transient Evoked Otoacoustic Emission Testing assess the otoacoustic emission from the ear when a sound is initially introduced to the ear. The information is used to determine the ability of the ear to send sound to the brain. Early on, there were three major applications for the testing of otoacoustic emissions:
1) Diagnosis of hearing loss;
2) Hearing screening in difficult to test patients such as newborns and
3) Monitoring of progressive hearing loss.

Few use the information from spontaneous otoacoustic emissions to determine issues relating to body specific wellness challenges.

There are two basic classes of otoacoustic emissions (OAEs): evoked and spontaneous. The Distortion Product and Transient Evoked are considered evoked. A response is evoked due to a stimulus being introduced to the ear. Also within this class are Stimulus Frequency

OAEs. The spontaneous OAE has not been used clinically for diagnostic purposes mainly because of the individually based uniqueness of responses emitted. Spontaneous OAEs are not evoked because they are always present and do not need a stimulus to capture them.

The concepts expressed in this book use only spontaneous otoacoustic emissions. These spontaneous OAEs are obtained by inserting a probe tip containing a special microphone into the person's ear canal. The emitted sound is captured by the probe tip, sampled and averaged over a brief period of time, and then is analyzed for spectral information. The frequencies captured are typically between 500 and 7000 Hz although more extensive ranges are emitted. The results are displayed in graphic formation. Otoacoustic emissions are known to be synchronous (or in patterns) and frequency specific (having specific frequency responses).[4] My research in The Davis Addendum to the Tomatis Effect connected irregular patterns in the spontaneous otoacoustic emission graphic formations with the vocal irregular patterns evidenced using voice spectral analysis.

Most researchers and audiologists are interested in otoacoustic emissions as a tool for testing the status of hearing function. What happens to the emission from the ear if there is difficulty with how one hears sound? Typically most otoacoustic emissions are weaker if the inner ear is defective, and not surprisingly, appear to decrease with age. The levels of sound emitted appear to be greater in children than adults. Children also seem to emit higher frequencies than adults.[5]

Not all people appear to emit spontaneous otoacoustic emissions, although with the improvement in the sensitivity of equipment more are being measured. Current thought is that seventy-five percent of all populations emit spontaneous otoacoustic emissions.[6] However, the absence of these emissions does not mean that the person has an auditory dysfunction. Aspirin, for example has been shown to inhibit spontaneous otoacoustic emissions.[7] ATP* receptors are considered to be important with receptor activation in various cell types in the cochlea.[8] If ATP is unbalanced, what happens to the otoacoustic emission? Further

---

* ATP is Adenosine-5-triphosphate and helps transport the chemical energy within cells for metabolism.

research is needed to expand our understanding of this important emission and how it impacts the body.

Neuroscientists at Johns Hopkins University in Baltimore, MD discovered how cells in the developing ear make their own noise before the ear is able to detect the sound around them. While studying the properties of non-nerve cells in young rats these support cells showed robust electrical activity without sound or external stimulus. They found that ATP was being released near cochlear hair cells responsible for sending sound to the nerves. This team showed spontaneous electrical activity occurring at the same time as supporting cells. While the study looked at young rats the hypothesis was suggested that "if ATP were released by the remaining support cells, it may cause the sensation of sound when there is none",[9] such as with tinnitus. How does this affect otoacoustic emissions? In order to answer that question more research needs to be done.

Most researchers are evaluating spontaneous otoacoustic emissions by the characteristics of their signal: the phase, frequency, intensity, duration of amplitude, suppression, etc. These researchers are evaluating the characteristics related to the hearing or auditory function. Perhaps an idea for future research would be to consider the signals related to their vibrational response. We know that the cochlea acts as an interface between the physical world of sound and the brain via neural transmission. Perhaps we should now consider that this world of sound emission is not for hearing alone, but is the sound for connecting the internal self with the outside world's energy vibration.

Vibration includes movement. The importance of movement within the body from cellular movements to movements of the entire body is integral to The Cycle of Sound. The inner ear has both inner and outer hair cells. The outer hair cells utilize a motor process whereas the inner hair cells invoke a sensory process. The movement process associated with the motor function of the outer hair cells agrees with the concept of the entire body being a system of movement patterns and rhythms. The motion of the hair cells is subatomic at threshold, and minute changes in sound can trigger a response. The hair cells of the inner ear can pick up very subtle differences in sound changes. Perhaps this means both receptively and expressively? These movement patterns continue to support the cyclical nature of the Voice-Ear-Brain Connection.

Recent research indicates that these inner ear hair cells operate like 'constantly moving treadmills',[10] moving two proteins from the bottom to the top of the hair cell. The one protein is called a motor protein because it burns energy to move the material within the cells. All of our energy processes will be in movements or cycles.

My initial interest with spontaneous otoacoustic emissions began when one of my clients introduced me to Sharry Edwards, the founder of BioAcoustics. I was fascinated that she could hear the sound emitted from the ear, and that she began to make sense of what these sounds meant by connecting them to the voice.

But then I had an *ah-ha* moment when I found Kirchhoff's Principle which says that the frequencies absorbed by a molecule are identical to the frequencies emitted by a molecule when excited. And that the energy is absorbed by the reverse of the process through which energy is emitted. The absorbed energy sets up movement within the molecule, forming a continuous energy system.[11] This *ah-ha* incident helped me understand what I needed to do when I researched and published about The Davis Addendum to the Tomatis Effect.

James Oschman in his numerous writings discusses the work of Herbert Frolich who suggested that crystalline molecular arrays can vibrate strongly and coherently. He presents two qualities that are important. One, that crystalline molecular arrays are sensitive to energy fields which some suggest are impossible. The second, that strong vibrations travel within the crystalline network of the body and also radiate into the environment. Researchers have detected vibrations of many frequencies including visible and near visible light frequencies that also appear to have biological effects on humans.[12] These crystalline molecular arrays within our cell network act as antennas emitting and receiving signals.

Perhaps the cilia of the outer hair cells act as this crystalline antenna. Currently, scientists are limiting the work with spontaneous otoacoustic emissions to hearing and auditory processing issues. Let's blend together the many sciences and look further. Perhaps our spontaneous otoacoustic emissions are informing us of our body's irregularities at the cellular level and as defined within the margins of the Voice-Ear-Brain Connection. Energy therapists using other modalities already know that the human body emits vibratory information about what is happening inside. This is a definite possibility for spontaneous otoacoustic emissions.

Ototoning

A new term is introduced: Ototoning. This newly coined word can be considered the pinnacle of thousands of years of sound work searching for the ultimate body enhancement and support system. The top of the mountain has been reached and we, as humans, can finally move forward. The ultimate personal sounding response has been found.

Ototoning: *Oto* refers to the ear and *toning* typically uses the repetition of a vowel to resonate throughout the body. Therefore, envision Ototoning as using the ear to determine which sound to resonate throughout the body. Ototoning uses the sound emitted from the ear, which the voice then tones back into the body. Ototoning can have a powerful effect on the user.

This new word simply means using the sound coming from our ear as the sound the body is saying it needs to self-heal, and then toning that sound back into the body to bring the body into harmony. This technique uses the power of the voice vibrating through the body to make major energy changes for the person using the technique correctly. This idea and technique will be discussed throughout the book but is based upon our ear's ability to make sound, not only to hear sound.

The mystic who understands that the secret of all knowledge must come from within the self, also knows that the value of the 'word', or their important personal sound, comes from within himself as well. Hazrat Inayat Khan shares:

> "What is sound? Is it something outside, or is it something
> within? The outward sound only becomes audible because

the sound within is going on, and when the sound within is shut off, then the body is not capable of hearing the outward sound. Today the human being has become so accustomed to external life that he hardly even thinks of sitting down alone, and when he is alone he occupies himself with a newspaper or something else. By always occupying himself with external life, a person loses his attachment to the life within. His life becomes superficial, and the result is nothing but disappointment. For in this world there is nothing in the form of sound, either visible or audible, that is as attractive as the sound within."[13]

Consider this quote as it relates to the Voice-Ear-Brain Connection. We need to hear the sound from within our body as emitted by our spontaneous otoacoustic emissions. We cannot feel alive if we do not hear the sound within. We must be able to hear the sound from within as expressed in the outward sound of the spontaneous otoacoustic emission. We can only tune into this sound in the silence (or with a device that will help you hear it). For some, this hearing is through our internal awareness of the vibration of the body, but this type of response is not as whole as physically hearing the frequency specific information about the sound. However, once we hear this sound we then want to produce that important sound back into our body by Ototoning. The sound produced as an outward expression of the voice is lifeless if we have not heard the sound from within. It is important for us to not turn off the internal sound by being so absorbed with the external world that we cannot hear it.

A poem by Elyse Betz Coulson reads:[14]

In the beginning
Was the stillness
Of All That Is.
The stillness moved
And there was sound.
The sound took form
And became the word.

> The word was God,
> The word is God,
> And then, the word was made flesh
> And the individual I was given control
> Under the law of the word.

I feel compelled to add the following to this beautiful poem:

> Yet it was within the silence
> That I found the meaning of the sound
> The sound that provided
> A more balanced I.

It is from within the silence that we find the correct sound to tone for our body to self-regulate itself. This is the concept of Ototoning. How can we maximize our use of Ototoning? Sharry Edwards has already created a matrix of mathematical connections from this Voice-Ear-Brain Connection. She has correlated the emissions from the ear with the voice and can introduce back into the body what it needs to repattern or self-heal. She is not claiming any medical changes. She only claims that she can support natural self-change. In the future I anticipate producing a device that will help make Ototoning a more frequency specific activity by collaborating with Ms. Edwards. Right now, my concept of Ototoning allows us to generically use the sounds we hear from our ears to be reintroduced to the body.

Considering this, the voice then becomes key to understanding our overall energy. John Beaulieu in his book *Music and Sound in the Healing Arts* says that "Speaking is the movement of Sacred Sound seeking expression…When speech is not in alignment with our inner self, then there is dissonance, which causes a dissipation of energy".[15] Since I have described this Sacred Sound as the sound emitted from our ears, one can restate this to say that, "The voice reflects what the ear is trying to express as out of balance. If the voice output is out of balance, the brain doesn't know how to maximize the overall body energy." This expression of the spontaneous otoacoustic emissions being connected to the voice was first identified in my Davis Addendum to the Tomatis

Effect. The voice reflects our overall energy but the ear is providing the missing piece for maximum energy. The emission from the ear provides information about the inner life energy. The voice reflects this energy. The brain then simply uses this connection to support the entire body. This then demonstrates the Voice-Ear-Brain Connection.

Silence

Silence is not the absence of sound. To be truly without sound would be like being in an anechoic chamber where after a period of time we get disoriented. Silence is being in a quiet place devoid of extraneous sound: TV, music, people talking, laughter, car noises. Silence is a place where you can hear your own body sounds; a place to tune into your own self.

Fabien Maman in his book, *The Role of Music in the Twenty-First Century* shares his experience of playing music for an audience. He found that the silence of the audience during each musical piece helped him create the music while playing. He began to tune into the resonant realm of the audience's silent music. He was inspired and moved by their response during this silence. For him, the silence was full of music. He was getting musical content and emotion from the silence between the musical notes.[16] He was able to provide the audience with the needed sound for the entire group by listening to the silence of the group.

Within the context of this book, silence is needed for the individual person – not the group – because each person needs to find his own sound within the silence surrounding his body. The sound of the person next to you or the sound of the heating unit in your room may interfere with being able to find the sound that your own body wants and needs. Only within a quiet environment will each person be able to find that sound. In effect, we are our own composers creating our own song and conducting our own orchestra. We need to feel balanced between the vibrations of the universe around us and the internal vibrations of our body.

Movement and the Ear

Body movement is so very important to any task that we undertake, especially when sound related. Ancient rituals used sound and movement for healing purposes. Sound was often in a chanting form and movement was often in a dance form. Our ancestors instinctively knew that movement needed to be combined with sound. Recent research has demonstrated repeatedly that movement makes you feel better, that music and movement helps prepare the body to be more focused and responsive to learning.

Movement and sound go together. Movement is an important adjunct for maximizing any sound program. The body is in constant cellular motion and the ear is the stabilizer of movement responses. After a study testing people with auditory synesthesia (where individuals hear sounds such as tapping or whirring when they see things move or flash), researchers at Caltech Brain Imaging Center in Pasadena, CA suggest that the motion processing centers within the visual cortex may be more interconnected with auditory brain regions than previously thought.[17] In this way the ear and the auditory centers of the brain are also involved with movement and are the starting points for triggering body responses.

In a recent study from Brunei University, a sports psychologist reported that training to music lowers your perception of effort and can trick the mind into feeling less fatigued during a workout. By matching the beat of music with the tempo of an exercise, participants could regulate their movement and reduce the oxygen needed to run by six per cent.[18] Also, specially chosen music enhanced physical endurance by fifteen percent.[19] A big industry has been launched producing CD's of fitness tracks for workout routines.

In a recent article about children with ADHD and movement, the author expressed the possibility that the hyperactivity of these children may actually increase alertness and facilitate their learning process. Mark Rapport, PhD at the University of Central Florida in Orlando demonstrated that children with ADHD need to move in order to maintain appropriate alertness for performing tasks that require working memory. This type of memory helps people hold information long enough to use it in the short term. He suggested many movement activities to help students with ADHD be able to maintain alertness for learning tasks.[20] His work supports the need to move in order to process incoming sound information when a deficit is present. One of the suggestions Dr. Rapport had was to chew gum. It is interesting to note that the nerves stimulated when chewing gum are closely connected to sensation through the ear.*

In October 2007, a study reported in USA Today that children who play vigorously for 20-40 minutes a day may be able to better organize their schoolwork, learn mathematics, and do class projects. Brain scans found more neural response in the frontal area of the brain, an area for executive function.[21] How many of us have reported that we feel more responsive when we exercise? The Voice-Ear-Brain Connection will be in better balance if we are moving (as with exercise) and responding to sound, simultaneously. And those who crave exercise or feel better with exercise are most likely demonstrating a need that the Voice-Ear-Brain Connection is not in balance.

Bruce Lipton in his eight-hour lecture on his book, *The Wisdom of Your Cells: How Your Beliefs Control Your Biology* discusses that the secret of life is movement because movement is simply protein changing shape. Cells are made up of protein building blocks. Every protein is determined by a sequence of amino acid cells assembled into parts that need to move, thereby creating a body function. If the function doesn't work correctly, then a disease is expressed as a symptom.[22] This type of movement is one of the aforementioned cycles, yet is a body pattern that keeps us alive. Most of us do not think of our body moving in this way.

Margaret N. H'doubler wrote in 1946, "Movement is so basic a part of being alive that it is quite likely to be taken too much for granted."[23] Most people just assume that movement is a part of their physical being. People often share that exercise enhances their well-being. If they don't exercise daily they don't feel as energized. Think of the businesses built up around exercise. You can buy all kinds of programs, videos, and exercise equipment to enhance your physical fitness. Yet people who are couch potatoes often complain that they have little energy. Movement makes our bodies sing and feel alive. We do take movement for granted though. People don't even think of the ear as your movement center. Nor do they see a connection between movement and the Voice-Ear-Brain Connection.

---

* For more information about the ear and the nervous system responses, read Chapter 3 in *Sound Bodies through Sound Therapy* by Dorinne S. Davis.

Movement is key to feeling good and keeping our body functioning. Remember that our bodies are made of moving atoms and molecules. These movements are vibrations which can be defined as sound movements. If our body is always moving with sound, what will happen if the sound stops or the sounds go out of tune? Perhaps people feel wonderful after exercise because they are activating the overall movement of energy within their body and as a result feel more in tune. The energy of the cellular movement makes better healthy connections with cells throughout the entire body. Exercise enables us to maintain our health and wellness because the necessary movement within the cells are activated sufficiently, stimulating an energy vibration that triggers positive responses within the whole body.

Movement isn't just a physical movement from one place to another. Movement also includes rhythms and patterns. When a baby first enters the world he doesn't try to speak. Instead, movement is more fundamental to his needs. While the sense of hearing is fully functioning at four and a half months in utero, so too is the fetus reported as moving at this time. Fetuses have been recorded to be jumping, moving their limbs, and averting their heads in response to sound at four and a half to five months in utero.[24] It is from movement that we understand the rhythm and flow of connected speech sounds, emotional tones of people's voices, the differences between pitch and intensity, and so much more. It is from rhythm and patterns that we are able to develop speech and language.

The movement process is fundamental to the Voice-Ear-Brain Connection because of how the ear receives and processes incoming sound. As mentioned previously, the ear is comprised of three parts. Two of these parts are important with regard to our body's movement. So as a person receives a sound to the cochlea, the same vibration is triggering these other two parts. Movement and sound responses are sent to the brain along the eighth nerve. The brain stores movement patterns as wave patterns.[25] Movement supports the reception of sound, and sound supports the integration of movement.

We know the body has five senses: taste, touch, smell, sight and hearing. But some say that we have a sixth sense: proprioception.

Proprioception lets us know about our bodies in time and space, meaning where are we positioned or located at that moment in time. As every sound we hear or make creates a sensory memory, and everything we see creates a sensory memory, so too does every movement or body response create a proprioceptive memory. These memories, based on information from the ear, become buried within our memory storage system in the brain, but when new situations arise we use these memories to make new judgments and decisions. When a child is first learning to ride a bike much practice is needed. As the activity is repeated, the body remembers what works. After awhile bike riding is a natural response. Also, a person can visualize himself riding a bike and mentally feel how their legs are pushing on the pedals or how their arms are pushing on the handlebars when turning. We have created this memory for the whole body to experience and we can recall the event just by thinking about it. The highest level of Ototoning incorporates this same recall of the sound sensation needed for the body to self-heal.

Even though the Lower Leaves and Branches of the Tree analogy incorporates auditory processing skills, this area also incorporates vestibular, movement, and proprioceptive responses, mainly because the vibration sent to the cochlea is also sent to the semi-circular canals and vestibule. Therefore, the sound-based therapies at the Lower Leaves and Branches also include sound and movement techniques such as Interactive Metronome.

Within the Eurhythmics method (a technique using sound, body, and movement) each new task begins with listening, and then walking is added followed by singing. The movements and positions of the group are carefully watched. Each activity is stored in the brain so that the sensations can be recalled at a later event. Dalcroze, the developer of this method, reported that when advanced piano students were asked to play a fugue, they were able to play the fugue even after the keyboard was scrambled because "they turned off the sound coming into their ears and listened to the sound in their memory".[26] They recalled the sensations from their earlier playing sessions.

We have two basic nerve receptors for these sensory experiences: kinesthetic and vestibular. Kinesthetic receptors are on the muscles, joints, and tendons; vestibular receptors are in the ear. Both of these receptors support the ability of the piano players with remembering how to play their music. Eurhythmics supports successive repetitions of stimulus, perception, storage, and retrieval which support the recall of events. Dalcroze said, "Music training should develop inner hearing – that is, the capacity for distinctly hearing music mentally as well as physically." [27] The body is recalling the kinesthetic and vestibular sensations as well as the musical patterns, beats, rhythms, intensities, and melodies. Dalcroze's goal was to "develop this capacity to recall vividly any musical sensation from within oneself".[28]

Once the ability to Ototone is learned and the necessary sounds are identified, the capacity to recall the sound sensation and project it through the body without making the sound – but with the same results – would be an ultimate goal.

We also need to introduce structural alignment within the idea of movement. Dr. Ida P. Rolf, who introduced the structural integration technique known as Rolfing, knew that misalignment of any part of the body affected the entire system, and that by restoring balance *vertically* she could address many clinical problems. She found that for any movement there is a pattern of compression and tension that creates precision, efficiency, and gracefulness of that movement. She connected structural, kinetic, and emotional stability within her process.

By balancing the Voice-Ear-Brain Connection through listening to sound-based therapies, many clients have realigned their skeleton, improved their balance, their gait and motor planning, released emotional blockages, and much more. These are similar results to those found through the process of Rolfing. Good structural alignment is necessary when incorporating The Listening Posture (discussed later in this book) within an Ototoning session.

This can be understood by looking at our connective tissue network. Connective tissue comprises many physical structures within the body such as tendons, ligaments around cartilage and joints, and the fibers in muscles. Coverings of the nerves, lymphatic system, and blood vessels are also included in this network. Any movement of the body is made by tension through this connective tissue. Remember, sound vibrations travel through the connecting tissues as well as through the nerves and circulatory system. The vibrations also stimulate the vestibular and motor areas of the ear. The system of the Voice-Ear-Brain Connection has to send the sound vibration to the entire body because of where and how the energy is sent. The health of our body is dependent upon the flow of energy that is sent throughout it. So movement, balance, coordination, and gait can all be improved if the system is working better. The Voice-Ear-Brain Connection is a system that must be improved for all parts to work.

Old Concepts and New Ideas

Instead of saying, 'out with the old and in with the new', a better phrase to use with the Voice-Ear-Brain Connection is, 'blend in the old to enhance the new'. All of the processes and steps that I have been discovering over the past many years have attempted to make use of the knowledge of my predecessors. With the information presented in this book we can now blend in the new ideas to make the best process available to all. What is the best way to do this with what we know now?

Sound Notes
Chapter Two, Summary Statements

- Sound-based therapy uses sound vibration with special equipment, specific programs, modified music, and/or specific tones and beats. The need for sound-based therapy is identified through appropriate testing.

- The Tree of Sound Enhancement Therapy is the developmental flow chart for the correct administration of any sound-based therapy.

- The Tree of Sound Enhancement Therapy has six parts:
  1) The Seed: Basal Body Rhythms
  2) The Root System: Sense of Hearing
  3) The Trunk: General Sound Processing Issues
  4) The Lower Leaves and Branches: Specific Auditory Processing Issues
  5) The Upper Leaves and Branches: Academic Skills
  6) The Head: Support and Maintenance of the Body

- The Diagnostic Evaluation for Therapy Protocol (DETP) follows The Tree of Sound Enhancement Therapy flow chart. The results demonstrate where the body has broken down and compensated within the automatic energy flow of the Voice-Ear-Brain Connection. The protocol suggests if, when, how long, and in what order any or all of the many different sound-based therapies can be appropriately used.

- The Davis Model of Sound Intervention incorporates three concepts:
  1) There are five laws supporting a connection between the voice, the ear, and the brain;
  2) Every cell gives out and takes in sound; and
  3) The ear is our global sensory processor. Once the Diagnostic Evaluation for Therapy Protocol identifies the correct sequence of sound-based therapies, change occurs with the understanding of these three concepts in the person's sound-based therapy programming.

- The Davis Addendum to the Tomatis Effect is the completion piece for taking the use of sound received and expressed by our body to our full, total body enhancement, and demonstrates The Cycle of Sound.

- Spontaneous otoacoustic emissions are sounds emitted from the ear which demonstrate frequencies that the body is missing or are imbalanced within the voice. These sounds appear to be frequencies the body wants restored in order to feel balanced.

- It is within silence that the body can hear, feel, and experience the sounds that it desires.

- Movement is key to feeling good and keeping our body functioning. Our movement center is in our ear.

CHAPTER 3

Systems and Basic Sound

*I*t cannot be emphasized enough that the Voice-Ear-Brain Connection is not about hearing. It is not about how we hear, or how we process auditory information, or even the effects of music. The Voice-Ear-Brain Connection represents a balanced vibrational sound energy system within our bodies. This Cycle of Sound takes us to a new beginning – a new way of viewing the term sound and the healing response of the body. The science behind how sound vibrational energy impacts our lives is based more on physics, biology, and chemistry than on hearing, speech, and/or language. The process encompasses the *whole body* so the total person must be evaluated – not just behavior, emotions, learning, development, anatomy, and so on. The process includes the many systems of the body networking within the physical body, as well as the systems that exist in the energy levels outside of the body. The Voice-Ear-Brain Connection uses sound vibration moving outward from the body as well as being received by the body, creating a cycle of responsiveness.

How was this system identified? Jeffrey Maitland in a SOMATICS article from 1980 says,

> "Strictly speaking every system is itself a system composed of systems, every system is a member of a system, and the forces between systems are themselves systems. To under-stand any one system requires understanding the living body as a whole and to understand the body as a whole requires

understanding each system… no one part of a system is more fundamental than any other."[1]

This is most likely why the field of energy and specifically sound-based therapy is so difficult for many to understand. We do not just address one organ or one sensory system, nor do we just address behaviors, or emotions, or physicality. We address the entire person with sound-based therapy. We also do not consider one sound-based therapy as the most important. We look at the total person as described within the interpretation of The Tree of Sound Enhancement Therapy.

The Voice-Ear-Brain Connection is a system of the body previously unrecognized. By introducing the Voice-Ear-Brain Connection as a newly identified system we are able to look at how the entire body works together within itself and without to the environment. Similar to the acupuncture meridian points as a system of the body, so too is the balance of the voice, the ear, and the brain a needed energy system of our body.

Systems and Vibrational Patterns of the Body

Edgar Cayce said, "Life in its manifestation is vibration".[2] Hazrat Inayat Khan says that, "All existing things that we see or hear, that we perceive, vibrate".[3] Vibration has motion, therefore all life is in motion. "Philosophy or science, mysticism or esotericism, will all agree on one point …, and that point is that behind the whole of creation, behind the whole of manifestation, if there is any subtle trace of life that can be found, it is motion, it is movement, it is vibration."[4] Harmony of this vibration is the only thing that unites life forms. The relation of one sound to another creates this harmony.

Our ear provides us with responses to many different sensory perceptions. According to Dr. Alfred Tomatis, the brain receives more stimuli from the ear than from any other organ. And the vibration of sound through the ear will stimulate all the senses. The ear transmits sound to the brain where it is organized and sorted into the signals sent to the body as energy. In turn, each organ and function within the body creates a vibration which helps it maintain its equilibrium. These vibrations allow the body to cooperate with its own self-healing.

Quantum physics took hold during the twentieth century. Quantum physics is the study of atomic and subatomic forces underlying our physical world. Motion is at the crux of our physical world. Quantum physics is also the physics of possibilities. The possibility exists that what happens within us can impact what is outside of us and vice versa. Quantum theory shows that all matter is in motion. The particles that we see are simply movements within our consciousness. Our thoughts can change our perceptions. We can choose what we bring into our physical understanding, yet our senses will respond automatically when there is an alteration of vibration around us.

Recently I went to a health spa and while there I met a woman who had an unusual experience. She went on a field trip at the spa and noticed a whirlwind or tornado-like activity of wind coming towards the bus. Because this whirlwind seemed unusual, she took a few pictures as this wind tunnel came towards her group. Looking at the image on her digital camera she did a double take. It appeared that the tornado-like activity captured in the camera showed the image of a man. She quickly took another picture and the man was even clearer this time. Subsequently, she printed the image. Now, while standing in front of this whirlwind no images could be detected. Yet, the man's shape was very distinct in the picture: a dark-skinned man wearing a straw-like hat, a jacket, and jeans. This incident draws attention to the energy around us, taking shape as only our consciousness informs us. However, the reality of the form was there in her subconscious and appeared when the energy was sufficiently captured by the camera's quick snapshot. Her senses were aware of the alteration of the vibration but her eye wasn't quick enough or conscious enough to formulate the physical matter of the man. The shapes and forms that we see every day are only because of the way our eye and brain process the speed of the energy formulating the shape or form and bringing it into our consciousness. I include this photo for you to view the results for yourself.

Photo credit Maria Harris

In nature, everything that seems solid is not really solid at the atomic level. Within each atom are subatomic particles that are separated by huge gaps. Each atom typically has a void of about 99.999 percent empty space.[5] But the motion within this space and the atom infer the quantum theory of the universe. These vibrations are frequencies that support our body and its connections to the rest of the universe. Our internal vibrations are searching to be harmonious within ourselves as well as with the external vibrations of the universe. Deepak Chopra expressed that our atomic empty space may be a rich field of silent intelligence exerting a powerful influence on all of us.[6] The Irish physicist John Bell found that "all objects and events in the cosmos are inter-connected with one another and respond to one another's changes of state".[7] This means that if our internal or external vibrations are disturbed, the vibrations around them will create a change to the surrounding areas. These disruptions create an imbalance in our energy patterns. Our cells communicate with each other and each cell talks to the rest of the body, resulting in an unknowable number of messages being sent to the trillions of cells in our body. Every cell in the human body can also hold onto an emotion. Can you see now how all cells know everything the body is experiencing?

A new term has been introduced recently: dirty electricity, which has been described as high frequency voltage transients on electrical wiring caused by an interruption of electrical current flow. [8] Dirty electricity is considered to be a new type of electromagnetic pollution affecting our health. Electromagnetic fields (EMF) and radiation from the numerous technology devices available today appear to have adverse responses within our bodies such as skin reactions, heart problems, psychological problems, and changes within the central nervous system. What is really happening? Simply consider the energy flow of any outside disturbance.

These external emitting frequencies break down to sound or frequency energy waves which will distort or disturb the balance of the frequencies within the Voice-Ear-Brain Connection right down to the cellular level of the person. All day long, while exposed to this constant bombardment of technological frequencies our bodies must struggle to balance our systems. Each one of us will benefit from a daily rebalancing of our frequencies. The easiest way to accomplish this is with a sound-based therapy or with Ototoning.

The Voice-Ear-Brain Connection is at its best when the body is in a state of homeostasis or a place of equilibrium. All systems within the body must move toward a balance of the energy contained within the system. Internal disruptions or external influences can affect this delicate balance at any time. The body is constantly working to restore itself to its natural state of balance. The body is working to adjust the energies so that the systems are harmoniously working together. When one part of the system is out of balance, the disruption is felt by the other parts of the system. The body tries to compensate for this disruption by creating a functioning, though imbalanced system.

A little known balancing system of the body happens in the brain. The brain likes to shift cerebral hemispheres every ninety minutes. In other words, it does not like to sustain one continuous activity with only one hemisphere beyond ninety minutes. If it does, then health problems have been known to occur. In his book, *The Rhythmic Language of Health and Disease*, Mark Rider says "stress can be caused by too much homeostasis",[9] meaning that if we do one primary activity throughout the day we may impact our health, i.e. get a headache. I've noticed for example, that when I write for a period of two hours or more that I ultimately get a headache. I must get up and move around or change tasks before beginning to write again.

Biologically, all cells of our body have a constant flow of elements in and out of the cell structure for survival. The cells respond to stimuli, repair themselves, acquire nutrients, give out waste, and grow. This process of homeostasis provides the balance that the body needs to maintain life. This constant flow in and out of the cells can also be seen as a give and take of vibrational energy. Each cell has a resonance that needs to be in harmony with its surrounding cells. Activities such as muscle contraction, nerve conduction, and glandular secretion also produce their own resonance. The Voice-Ear-Brain Connection has the ability to bring homeostasis, working from the cellular level to bring harmony with the surrounding cells as well as the entire body.

Most quantum theorists believe that the world is a projection of invisible reality and that all things are not matter as typically envisioned. Quantum physics tries to measure things in the smallest unit possible. The proton was thought to be the smallest unit at one time, but the quark currently holds that distinction. It is a unit that one cannot see or touch. Some theorists propose that the building block of a quark is vibration, which has the potential to turn itself into matter.[10] Eddington, an astrophysicist of the early twentieth century, instead believed that the world is a formation of brain impulses.[11] Impulses travel up and down the nerves, which come from the vibrations of energy at the ends of the nerves. The foundation of this energy is an empty void, or a quantum void. The pathway formed is simply a code turning vibrations into experiences from which we determine meaning. These codes extend beyond our senses, but create an orderly reality for our existence. The visible universe is then simply a set of signals. These signals or vibrations can be compared to the subtleties expressed in the connections within the Voice-Ear-Brain Connection. The meaningless vibration can create a meaningful response in the code that is sent or received. The brain interprets the information, thereby providing meaning for the person.

Consider sound patterns, frequencies, or music within the body in this context. The impulses are produced, and then sent to the brain and the body, but they are just codes. One has to think outside of the body hearing the sound, and picture these coded vibrations producing invisible chords or tones without ever being physically heard. The body hears the sound with the quantum vibrations.

With the introduction of sound-based therapies the body's vibrations are being repatterned. Whether they are heard or not heard, the body is constantly working to establish a balance among its vibrational patterns. This balance is called coherence. Our limitations as humans keep us from perceiving these subtle changes that we make constantly.

Sound-based therapies simply provide a way for people to move into that balanced position with support.

Side effects have been reported with using sound-based therapies, but these reactions are simply the body's attempt to regain equilibrium. Everyone is different and some people have few or no recognized side effects while others have major debilitating ones. Some people have immediate positive change while for others the effects are taking place more at the core level and the positive outcomes are not known until later. And sometimes the outcomes are never known because the imbalance was addressed and eliminated before manifesting itself.

Some people regain a balanced or coherent system by using the sound-based therapies, but later notice that they have good and bad days, and can't understand why. When asked to track the weather for example, certain patterns arise. Because everything has a rhythm, when the environmental rhythms are disturbed with rainstorms, lightning, snowstorms, tornadoes, sunspot activity, or other geophysical or celestial distortions, these disruptions may impact certain sensitive people.

Our energy frequencies spread out around us and within us. Each wave meets up with other waves, changing the pattern and vibration of the new wave. Imagine a drop of water going into a pool. The water doesn't really move. The energy of the cyclical waves moves outward until they reach another droplet. The energy wave then changes course and gives shape to new energy and frequency. If a cork is dropped into a pool of water the cork moves up and down but doesn't move outward. The energy of the wave made when the cork is dropped moves through the water, not the cork itself. However, if red dye is dropped into a pool of water, the red dye dissipates until it is hardly visible as it blends with the other frequencies and waves in the pool. Our lives are compilations of many interacting waves, activities, and energies blending together.

Everything moves in cycles, and cycles within cycles. Every event influences another event so that cycles are always happening. Nothing is inactive. Immobility is only a state of matter reaching matched speeds. So then, when do we as humans feel balanced? Our human consciousness cycles between our connection with the sounds of the universe and the tiniest sounds from our cells. We are constantly searching for our self. Who am I? What am I? How do I exist? Why am I here? Our physical body is defined by its vibrating structure but our conscious and subconscious states of mind extend inward to our core vibratory structures and outward vibrationally to the universe. Our bodies heal with the communication within our cycles. The Cycle of Sound appears to be a major cycle for our sense of self, for being one with who we are and how we fit into the world around us.

Our cells have cyclical patterns. Bruce Lipton in his eight hour lecture on *The Wisdom of Your Cells* says that, "Knowledge of cells is knowledge of self-power".[12] He relates that our fifty trillion cells are in constant communication so that health is when the communication is in harmony.[13] Every cell has every pattern, structure, or behavior of every other cell, so that the person and cell are interchangeable. Every cell has a membrane or skin, and can breathe, digest, excrete, and more; the same goes for the whole person. The cell and person get feedback from its environment. Who we are is determined by the environmental signals transferred through and received by the cells. These signals come from the vibrations around them. But the membrane perceives the signals and translates them into a use function. When the perceptions are not accurate the body's functions do not work correctly.

Whole Body Subtle System

The Tree of Sound Enhancement Therapy when used for the correct administration of any sound-based therapy suggests the importance of the whole person – both the input and expression of sound from our bodies. Edgar Cayce once said, "We must change the vibrations if we would have the permanent relief for the body."[14] Only by supporting both the give and take of The Cycle of Sound for the body and making it stable can permanent relief occur. One cannot only work on the Tree portion – the Roots, Seed, Trunk, and Leaves and Branches; one must work also with the Head surrounding the Tree. Together, balance is achieved and the results can become permanent.

In Vajrayana Buddhism there is no division between mind and matter. Within this tradition matter in its subtlest form is a vital energy that is inseparable from consciousness. It is a mind/body connection or in this case, a brain, ear, and voice connection. Within this connection an intimate relationship exists between the energies within our body and the energies outside of our body.

Deena Zalkind Spear begins her book *Ears of the Angels* with the quote, "If it vibrates, it can be tuned. Everything is energy – violins, animals, people, potato chips, thoughts, feeling, and events. They all vibrate."[15] Ms. Spear learned how to change the silent sound of people and animals by recognizing an acoustical transformation with an inner sense that she had. She felt the sound. Perhaps she was tuning in to the spontaneous otoacoustic emission of the person or animal and helped them transform their inner vibrations to more multiple dimensions.

Five elements of energy movement have been used by many cultures: ether, air, fire, water, and earth. All have a primary energy often referred to as sound. The five elements derive as subdivisions. Within the Ayruvedic tradition the world was formed from a cosmic vibration as a soundless sound, *Aum*. The first element to subdivide from this cosmic vibration was ether which went on to create air, fire, water, and then earth. In Greek mythology ether was considered to be the essence of space. Sir Isaac Newton used the word ether to mean the invisible substance permeating the universe. In the nineteenth century, James Clerk Maxwell said that ether was "a material substance... supposed to exist in those parts of space which are apparently empty."[16] Einstein said that ether was necessary for the laws of physics to exist. This etheric level is important for understanding that an energy exists that helps connect everything in the universe.

Interestingly, in relation to the body the etheric center is the voice, incorporating hearing and the throat. First was the sound and then came the use of the voice![17] At this etheric level sound is important and is directly associated with hearing. Only one sense is opened at this level: hearing. But the main association for the etheric level is the throat because it can express the other senses which are only opened within the other four elements. How fascinating that for hundreds of years people have understood through the introduction of other processes and systems that there was a very important connection between the voice, the ear, and the brain. Yet it wasn't until The Davis Addendum to the

Tomatis Effect demonstrated the connection between the voice, the ear, and the brain that the actual system could be identified.

Interestingly, Don Campbell in his book *The Roar of Silence: Healing Powers of Breath, Tone & Music* connects the four elements that developed from ether – air, fire, water, and earth – to audition. Air, in the form of auditory energy from outside of the head travels to the middle ear where bones which are of the earth move the sound into the cochlea which is water. From there, electrical energy in the form of electrical charges as fire sends the sound to the brain centers.[18] Adding ether to Mr. Campbell's four elements connection to audition is necessary because ether is the fifth element encompassing all four elements and is the primary element of sound. Audition is understood as the ear, and sound as the voice. The brain then reconnects the sound between the voice and the ear to have the full Voice-Ear-Brain Connection.

Vibration is synonymous with frequency. Different frequencies of energy are simply varying rates of vibration. Matter and energy are different manifestations of the same substance including our physical and subtle bodies. The vibrational rate of the energy determines the density of the matter's expression of substance, i.e. the man within the whirlwind. Matter vibrating at a slow frequency is thought of as physical matter but when the vibration exceeds the speed of light, it is thought of as subtle matter. Subtle matter then is dense matter just vibrating at a faster speed.[19] Perhaps another way of expressing this is to say, "The physical body vibrates, but at a slower rate than its energy field. Vibrations in the energy field affect the vibrations of the physical body."[20] Here is a great quote by James Oschman: "The body has no shielding to prevent communication signals from leaking into the environment".[21] This also works in reverse as we are a constant give and take of communication signals by the sound energy vibrations that we create and receive. And even though the energy may not be audible to the human ear sound is still emitted by the waveform.

Donna Eden in her book, *Energy Medicine: Balancing Your Body's Energies for Optimal Health, Joy, and Vitality* says that "matter follows energy. That is a fundamental law of energy medicine. When your energies are vibrant, so is your body."[22] Energy comes first and matter is a manifestation of the energy. She talks of three combinations of energy pertinent to energy medicine: electrical, electromagnetic, and subtle. The electrical involves the movement of the electricity in the cell – every cell stores and emits electricity. As mentioned earlier, perhaps this electrical system may actually be a sound movement system as hypothesized by researchers at the University of Copenhagen. The electromagnetic energy moves in waves but is absorbed by matter as a particle. The frequency range is very large and much is still being learned about this type of energy because the concept is involved in quantum physics. The subtle energies are hardest to describe because the effects are felt but often cannot be measured or detected directly.

Dr. Ibrahim Karim in advancing the field of BioGeometry, shared that:
"The human body has both peripheral and central energy systems. The peripheral energy system creates the energetic boundary around the body; it includes the fat layers of the body that have the capacity to store information with the outside world. Via acupuncture points and energy pathways it communicates this information to the central energy system. From this perspective, the acupuncture channels and points serve to connect the internal energy of the body with the environment, receiving information about the environment so that the core energies can adapt to changing conditions."[23]

To connect both the peripheral and central energy systems, resonance is important. The vibration of things, or geometric forms that are similar will vibrationally resonate together. There are multiple movements of energy throughout the internal body systems. Every organ of the body has a set pattern of energy movement related to its specific function. These patterns are geometric in shape and support healthy movement for that organ. If the correct geometry within each organ is necessary for wellness, someday I will explore the geometric shape inherent in the connection between the voice, the ear, and the brain. It must exist!

Resonance has two basic categories: free and forced. Free resonance is what happens when an object begins to vibrate when placed near another object of the exact frequency. Forced resonance is when a vibrating object of a particular frequency produces a vibration in another object of a different frequency causing the second object to change its frequency. This second type causes entrainment of the frequencies or the one frequency changing to the other and typically stronger frequency. Entrainment is an important response when discussing how sound impacts the body. We want to entrain the weaker frequencies to the stronger responses in order to balance the responses.

The possibility of entrainment can impact us in a variety of ways. We often take for granted our decisions believing they come from informative choices. We can use the example of the volatility of the stock market to illustrate this. In a recent abstract by Philip Maymin of the Polytechnic Institute of NYU, music and finance are linked. He analyzed more than five thousand hit songs and fifty years worth of stock market variability and found an inverse correlation between the volatility of the stock market and which music was a hit at the time as related to a frenetic or steady beat.[24] During a volatile period the top songs have a low beat variance, and while in a stable period music showing a high beat variance is at the top of the charts. Everything is in cycles for our bodies, which means we will come out of the highly volatile cycle once our bodies are entrained to the more balanced frequencies of not only the music, but also the sounds around us.

Resonance may be the key for understanding the bioenergetic levels of the body. What does the word *resonate* mean? Resonate means to return to the sonic pattern. Resonance supports the body returning to harmonic sound patterns. The subtle energy at the etheric level (the first level outside of the physical level) is at a higher octave than the physical body. However, our entire body must work in unison, combining lower and higher frequency energies for what can be called our Signature Symphony of Sound.

Tesla says:

"There are harmonies and rhythms which permeate all of creation. This idea is as fundamental to ordinary mathematics as it is to electricity. There are octaves of energy, definite waves and rhythms that can be measured, frequencies and amplitudes and so on. From these simple elements are produced an almost unlimited number of variations.... Because there are various octaves of energy in creation, there are subtle counterparts to everything existing in the physical octave... By applying a charge of external energy to a relatively closed system, you can selectively energize a given octave of energy...It's a basic principle of resonance."[25]

Resonance has also been linked to the Law of Similars. Dr. Richard Gilbert in his book *Egyptian and European Energy Work: Reclaiming the Ancient Science of Spiritual Vibration,* shares the work of Chaumery and de Belizal, two French researchers. Their Law of Similars says that similar shapes are connected with each other by shape-caused waves, or the vibrations emitted by all shapes. Basically, "all things vibrate with precise energies which create waveforms that are not perceptible to the five senses, nor in many cases, to electro-magnetic detection instruments." [26] This linkage is not limited to three-dimensional shapes. Mental images also apply. Dr. Ibrahim Karim, the forerunner in the field of Radiesthesia who studied Chaumery and de Belizal's work, feels that all living things are in resonance with the other things in their environment to which they are linked vibrationally.[27] Resonance with the greater world comes from our activity, emotions, energy, and consciousness. This resonance sustains us or destroys us, keeps us well or makes us ill.

As explained in the first chapter, energy of different frequencies can coexist within the same space without destructive interaction. The energy is invisible to the eye because it exists at a threshold of energy beyond the frequency sensitivity of our physical sensory organs. The difference between physical and etheric matter is frequency. The etheric body has a flow of energy exchanged through specific channels to the physical body's systems. Because of this, both physical and etheric matter can coexist in the same space. These systems combine to form an extended energy network of the etheric and physical planes. One such known system is the acupuncture meridian system.

This type of system which uses the principle of matter of different frequencies interconnected within the physical body may help support the concept that the Voice-Ear-Brain Connection as with the acupuncture meridian system is an undefined yet important system of the body. The channels of energy shared within the Voice-Ear-Brain Connection and the physical body support the entire body towards a balanced whole. As this concept is new with this writing, exploration into this connection is important for the future in understanding how we learn, develop, and maintain our wellness.

In the 1960's in Korea, early studies by a team of researchers lead by Professor Kim Bong Han found a connection between the acupuncture meridians and the nuclei of cells.[28] Because we now know that cells give out and take in sound, and that a connection by frequency has been identified for each cell and/or part of the body, the subtle energy flow of information is being channeled to the brain and back to the body by this flow of sound energy. This energy in and of itself is a subtle energy system within the body. Because this energy works with the laws expressed within the Voice-Ear-Brain Connection, the Voice-Ear-Brain Connection can be considered its own system within the body. Perhaps one day it can be viewed in connection with the nervous system, circulatory system, and soft tissue network which are already supporting the transportation of sound energy throughout the body.

Within our body's subtle energy systems the chakras also appear to be channeling higher energy to the cellular structures of the body. There are seven major chakras in the body and each is associated with a major nerve plexus or major endocrine gland. There seems to be an extensive network of fluid-like energies paralleling the body's nerves.[29] The energy flow of the chakras appears to be flowing in two different directions: one energy flow entering the body and the other exiting the body. This is similar to The Cycle of Sound: sound is received by the body and sound exits the body.

In his book *Vibrational Medicine*, Dr. Richard Gerber discusses five levels of the human frequency spectrum. The physical body is the most basic and understandable level because the energy can be seen and felt in what is typically known as matter or sensory received information. He describes the second spectrum or octave as the etheric body, a pattern that directs our cells from an energetic matrix at a higher level.[30] We are surrounded by an energy template guiding our body and life. This includes a give and take of vibration. We are sometimes influenced by the outside sources of vibration which in turn go to our cell structures. In other words, these outer vibrational influences can direct our cellular response. Illnesses sometimes can start in the etheric matrix before manifesting themselves at the cellular level. Rupert Sheldrake suggested that a pattern first appears in our subtle body before it becomes dense enough to imprint its message on our cells. The physical body then simply duplicates the pattern that resides in the etheric body (or higher).[31] Our mental processes are also a part of the etheric body. Our physical and mental health depends upon the energy at the etheric or higher body frequency spectrums. We must be able to find the correct frequency or combinations of frequencies as related to the sound-based therapies on the Tree analogy to support the well-balanced body.

The third level of the human frequency spectrum as discussed by Dr. Gerber is the astral level where our emotions are expressed. At this level a person's emotional state can disturb or enhance their health.[32] Western medicine looks to the neural activity of the limbic system within the brain for emotional expression. The brain actually helps with the expression of soul and emotional reactions into one's physical life. If the person's nervous system is impaired with disease the emotional response can be trapped at this astral level. The emotions congeal and increase in intensity over time if not released. When not released, these emotional reactions reach the cellular level and eventually vibrate through the Voice-Ear-Brain Connection. This vibration of sound energy via the pathway to the brain also passes through the limbic system. When sound-based therapies are used, the held emotional responses experience a vibrational release, or breaking apart. Perhaps this is why some sound-based therapies have a major emotional shift in energy for the person.

The fourth level of the human frequency spectrum is the mental body. This level, as with all levels above the physical body incorporates matter of a higher frequency. This level helps with the expression of human emotions, manifesting self and expressing intellect. Creation and transmission of concrete thoughts are important at this level. Above this level is the causal level which incorporates abstract ideas and concepts. This level deals with the essence and underlying nature of any concept. When supporting someone with an issue, starting at the causal level and backtracking through the lower levels will energetically involve all levels of the human frequency spectrum for a better release of the issue.[33]

Dr. William Tiller, PhD is one of the leading theorists in subtle energy. He designed a model for understanding the human energy spectrum and included more defined sublevels within the mental body, and put the causal body as the spiritual mind. He expressed that the interface between the physical and etheric bodies is influenced by the higher ordered energy connections.[34] Our physical senses and our consciousness make us think the world is as it appears. But we often don't see the unseen levels, processes, and interactions of subtle energy matter. As quantum physics has demonstrated, the true nature of reality is beyond our ordinary sensory capabilities. Dr. Richard Gerber said, "Our various subtle bodies would appear to have evolved for some purpose aside from the maintenance of the physiological functions of the physical body... The subtle energetic fields precede and organize the formation of the physical form as a vehicle of expression for higher conscious energies."[35]

The voice, ear, and brain are the physical form needed to share the cycle of incoming and outgoing sound stability to the body. This stability provides a necessary balance for each person. This subtle energy as stated above, precedes the final outcome. The precession of sound energy within the Voice-Ear-Brain Connection allows for integration of the energy so that balance can occur. The hypothesis is that the Voice-Ear-Brain Connection bridges the gap between the physical and the etheric levels, thereby bringing the physical to a more natural starting place for learning, development, and wellness. Along the descending energy levels from causal to physical, we must release the abstract, the intellect/sense of self, the emotions, and also release the energy of the higher frequency systems level, in order to reach the physical form integrating the sound energy cycle. It is this connection with which we experience and integrate our response and expression of sound. Whereas "the acupuncture meridian system is an interface of energetic exchange between our physical body and the energy fields that surround us", [36] so too is the Voice-Ear-Brain Connection an interface of energetic exchange between the physical body and the vibrational sound energy within the surrounding environment.

There is a fundamental dependent nature in reality which in Buddhism is called dependent origination. This principle is described in three ways:

> "First, all conditioned things and events in the world come into being only as a result of the interaction of causes and conditions. They don't just arise from nowhere, fully formed. Second, there is mutual dependence between parts and the whole; without parts there can be no whole, without a whole it makes no sense to speak of parts. This interdependence of parts and the whole applies in both spatial and temporal terms. Third, anything that exists and has an identity does so only within the total network of everything that has a possible or potential relation to it. No phenomenon exists with an independent or intrinsic identity." [37]

This principle helps in explaining the complexities of the Voice-Ear-Brain Connection. It is inherent in each one of us.

Once the ear begins forming in utero and subsequently develops further, the interaction of vibrational energy to and through the ear, the bone structure, the nervous system, the circulatory system, and through our cellular structure lays the foundation for the conditions by which our Voice-Ear-Brain Connection functions. This newly discussed body system needs time to establish its connections once begun in utero and with life development. There is a mutual dependency between the parts of the connection – the voice, ear, and brain – and the whole connection.

Without the parts of the connection, there can be no whole. The interdependence applies to time and space concerns as well as the expression and reception of sound energy. And lastly, the Voice-Ear-Brain Connection exists only within the system inherent in each one of us, but yet must work in compliance with the energies of the rest of the body, the sharing and reciprocity of the cyclical energy of sound. Each part of the Voice-Ear-Brain Connection must work in tandem with the other although each part can work and be studied independently. Within the Voice-Ear-Brain Connection system the parts work more powerfully together because of the connective energy flowing between and among the parts. This also is appropriate for the various sound-based therapies. Sound is powerful and each sound-based therapy can make change. However, the therapies work best when the connective energies flow as when administered through the wisdom of The Tree of Sound Enhancement Therapy.

Sound Intervention

In India, the saying, 'Nada Brahman' means that the world is sound! People have intuitively recognized that sound is an inherent, important quality not only within us but also outside of our being.

Pythagoras was one of the first to write about musical harmony. He expressed that all nature is in harmony as a result of the ratios of simple number frequencies. He experimented with hammers, stringed instruments, and wind instruments. The simple ratios between the size and weight of implements make for more pleasant sounds. They also provide the basis for our musical scales.

We have almost eleven octaves within our human range of sound. The lowest frequency we can typically hear is 16 cycles per second or Hertz (Hz) and the highest is 20,000 Hz. Below 16 Hz, we experience frequency as rhythm. Our brain seems to enjoy the relationship between simple harmonies, but when the harmonics increase in complexity the brain considers the sound as unpleasant. Within the octaves, natural patterns evolve so that whole tones and half tones are present.

In 1787 Ernst Chladni took sand, scattered it on a square plate and vibrated it with certain frequency notes. The sand formed different patterns on the plate. Harmonic tones produced nice shapes and patterns, but non-harmonic tones created confusion.

The Vedic rishis perceived a cosmos comprised of strings called sutras or verbal phrases. Billions and billions of these phrases were woven throughout the universe; therefore, the fundamental level of the universe was made of sound. However, more than one sound was needed to make the universe. And one sound broke the silence of the cosmos. This sound vibration was called *Aum* or *Om*, simply the first sound to break the barrier. This sound then divided into smaller waves and different sub-frequencies which then created the matter and energy of the universe.[38]

In the anonymous poem, *I Am Music* the author says, "Even the deaf hear me if they but listen to the voices of their own souls." [39] When someone is deaf, it is his or her sense of hearing that is affected. They still respond to sound vibration both within and without their body. If we refer to the sound coming from their ear as related to their soul, then they too can find the sound with which to Ototone. Most researchers have said that the deaf do not produce an otoacoustic emission; yet, those researchers do not differentiate between the natural emission (spontaneous) and the designed emission (evoked) typically used for testing cochlear hair cell response. Perhaps their emission is present yet in a different format, such as a non-hearing frequency? The deaf also have difficulty matching a pitch that they cannot hear so a future adaptation of the Ototoner™* will probably be a better match for them.

Civilizations and cultures have been known to use the personal sounds of their people for their enhancement or to support them in their life journey. An ancient Egyptian belief states, "The soul is in the tongue: the tongue is a rudder or steering-oar with which a man steers his course through the world."[40] It's interesting that they considered the tongue as the rudder since the tongue is important in creating the sound of the Ototone for the person, and the tongue is used for articulating the sounds of the voice – the indicator of our wellness.

---

* Ototoner™ is a trademark of Dorinne S. Davis.

In old Peru, the natives used Peruvian whistling vessels as a way to expand innate human sensing abilities. The experience resulted from the blending of several senses so that they were perceived as a single new sense. The sound was not music per se, but instead "a special toning of whistles which, when blown together with other such vessels, had the ability to induce what has been described as a 'trance state of a spiritual nature'. The sound was produced by a clay instrument, which was such a personally treasured item to its owner that it was not passed on to another person when he or she died, but followed the owner to the grave."[41] The vessels were finely tuned for each Peruvian tribal culture because each tribal culture had its own primary frequency. The process was wiped out when the Conquistadores invaded the country.

The Reverend Ted Karpf in the foreword to Don Campbell's book, *Sound Spirit: Pathway to Faith,* relates another cultural story about the aboriginal people of Australia. Each person had a melody that guided him or her towards his or her own songline this is described as a grid covering the world. It was felt that once the person found their song and their songline they could never get lost.[42]

In Roman times, the philosopher Boethius set out four classifications of music. One of the classifications, Musical Humana dealt with the rhythms of the body creating a harmonious resonance between the soul and the body. Illness was present when the connection became dyssynchronous.[43]

Historically, there are numerous mentions of the importance of sound or its components as a connection to the human soul and a person's well-being. Many have realized how to incorporate pieces of the Voice-Ear-Brain Connection, but did not necessarily know that was what they were doing. Probably the form most aligned with The Cycle of Sound comes mainly from the many different practices of yoga.

Sound-based therapies can make change for anyone, but the changes are more clearly recognized when placed along the continuum of sound-based therapies within The Tree of Sound Enhancement Therapy. Presently, there are position papers by important organizations in the United States against these therapies. Unfortunately, these responses only demonstrate the ignorance of those who have reviewed the information to date and those who are creating therapies that they insist are independent of each other. Mainly the problem has arisen in what they are testing and researching. Sound-based therapy uses sound vibrational energy to make change for our entire system but specifically in how the Voice-Ear-Brain Connection is functioning.

As stated earlier, many of my clients or potential clients think of my work as related to hearing or auditory processing. However, I work with sound. By using the word 'sound' with the word 'therapy' others conjure up their own idea of what that means. People think The Davis Center uses music or a form of music therapy. Others envision the center as using a bunch of CD's and having people listen to them. In fact, today's concept of sound therapy is often seen as people listening to musical selections and responding. Often, I'm asked to suggest a musical CD for listening. Our society typically considers sound therapy as passive listening to music, but sound therapy involves much more than listening to organized patterns of vibration frequently known as music. We need to actively involve the movement of and within the body, and especially use our voice to feel stimulated with sound. The research literature abounds with articles demonstrating the powerful response people (and animals) can have with music. People listen to music to feel better emotionally and physically. Although music therapy started after World War II the field has finally advanced to its currently recognizable professional field.*

Sound therapy can encompass musical selections that are modified or adapted specifically for the person, but sound therapy can also incorporate specific frequencies or specific tonal patterns to make change for the individual. Sometimes people picture me performing rituals with chants, or drums to make change. So recently I defined a few words. Their definitions may help you understand how I view my work.

- *Sound Healing:* Using tones, mantras, various rhythms and sound patterns, and other sounding techniques to support change with the body's energy patterns;

---

* For additional information about music therapy, visit http://www.musictherapy.org/research.html.

- *Sound Therapy:* Using specific notes of the musical scales to make change with the body's energy patterns, usually with a sounding source;
- *Sound-based therapy:* Using sound vibration with special equipment, specific programs, modified music, and/or specific tones/beats, the need for which is identified with appropriate testing;
- *Music:* The Artwork of Sound;[44]
- *Music Therapy:* Using the Artwork of Sound to make change with emotional, behavioral, and physiological body responses.

Music being the Artwork of Sound allows you to picture a beautiful painting of various notes all blended together to create a sound masterpiece. Listening to that masterpiece can indeed have a great impact on the person, but music therapy takes that masterpiece and uses it to accomplish a meaningful goal.

I view the world of sound intervention as having the above-mentioned three parts: sound healing, sound therapy, and sound-based therapy. In the past, I have specialized in sound-based therapy, but with the introduction of Ototoning I am bridging the gap and have now entered the world of sound healing and sound therapy. My work is now fully encapsulated with all components of sound intervention.

Sound Properties

If we think of sound as vibration, then anything sound comes in contact with will literally move. So when we hear sound, it moves us, especially the portions of the ear that trigger the response to frequency. When we feel sound, it moves us. We feel this sound through our bone structure and cell structure. When we sing or speak, we set a movement pattern in motion. When we listen, our body responds to the movement patterns received. Sound patterns are changed by the medium through which it is being transmitted. The molecules of the medium react to the sound vibrations in different ways.

There are four qualities of sound which, from an acoustical perspective, we judge as a sound. These qualities are pitch, timbre, loudness, and richness. Humans respond to these subjective responses.

Pitch refers to how humans subjectively perceive frequency: in cycles per second or Hertz. Pitch can be low or high-pitched. It is said that the ear can hear sound frequencies between 18 to 20,000 cycles per second. When these specific frequencies are emitted, an overtone is produced, meaning that the harmonics of the sound provide a more enhanced perception of the sound for a richer, fuller tone. To some, the overtones are like an unconscious radiating field surrounding a specific frequency, unless the sound has been produced as a pure sine wave, or single electronically emitted frequency. The specific frequency is known as the fundamental tone which we consciously hear. The overtones are produced in a mathematical relationship with the fundamental tone.

Timbre relates to our subjective ability to process the various combinations of overtones. The combinations of the mathematical relationships of the specific pitches or frequencies allow our mind to determine if we like the various combinations of overtones. The timbre helps us determine if we consciously hear just the pitch or the combinations of overtone notes within the specific tone. Timbre provides the subjective sensation that enhances sound in our mind for a more meaningful response.

Loudness is the subjective response to intensity – how much energy is needed to make the sound. We measure intensity in decibels (dB), a logarithmic change in our perception of the loudness of sound. Our ear has the capability to hear minute differences between sounds. We can consciously notice a difference in loudness of three decibels. Unconsciously, our body responds to fractions of decibels or extremely sensitive loudness differences. These responses are often subconsciously interpreted by the body, but we use and integrate these small differences all the time.

Richness is the subjective quality of the density of a sound. Richness refers to the quality of a sound, or the quality of the energy of the sound in space. Sometimes there is a lack of richness to the tone of the sound, so the sound is thought of as being too thin, or lacking enough density to provide the listener with a wonderfully powerful sonic environment.

Good singers and vocalists need to consider these four subjective qualities since they want to make the best sound to stimulate the listener. When Ototoning we can consider qualifying these four categories. We need to recognize that a specific frequency is needed by the body, we need to be able to match the pitch, and then produce the overtones for that frequency. Loudness is not important as an expression of outbound sound for Ototoning. The tone created is quietly expressed outwardly but can be powerfully expressed internally through bone conduction vibration for a louder, richer sensation of sound. The full complement of these four subjective qualities is demonstrated once the technique of using the correct Voice-Ear-Brain Connection vibrations or Ototoning is internalized.

Another way sound moves us is emotionally. Sound-based therapies have stimulated change with emotional responses. As previously mentioned, on the pathway to the brain vibrational sound stimuli pass through the limbic system of the brain. This section of the brain appears to be associated with the functioning of the autonomic nervous system. Our heart rate and breath stream are subconscious autonomic processes which can be controlled by the conscious mind. The emotional pieces are often more difficult to consciously repattern. Viewed this way, consciousness is a form of energy, but it is the highest form of energy at the physical level and very involved with our life process. The emotional pieces are often more difficult to consciously repattern due to our inability to reach that highest form of energy. This higher-level input influences the cells of the physical body, thereby helping to maintain order and balance at our molecular level. To maintain our health and wellness there must be balance between the physical and higher dimensional systems of the body at the cellular level. When there is a disturbance at either level, pathologic changes can be seen. But so, too, can emotional changes occur.

Our thoughts and emotions are manifestations of our higher energy vibrations. Emotional problems can be seen as energy imbalances affecting both the physical and subtle body response levels. So if the emotional problem has created a disturbance in the person's energy fields, sound-based therapy can support the person toward positive change.

Why only the Voice, Ear, and Brain?

Many healing modalities stress the air or breath stream as being key to success. Understanding the importance of the breath is essential for our ability to produce speech, for our ability to maintain our brain cells, for our ability to cleanse our cells, and for its importance to our rhythm in life. However, I needed to evaluate the system of the Voice-Ear-Brain Connection. This system begins and ends with sound vibration, both the reception and expression of sound. Every cell in the body has its own sound. Even inanimate objects have their own sound. Everything has sound and breath is not needed for this sound to be produced. Breath is used for vocalizing sound and maintaining our physical life. The rhythm of our breath is included at the Seed level of The Tree of Sound Enhancement Therapy because the seed is formed and starts to grow when combined with all of our body rhythms.

There appears to be a natural flow of energy in our body that will benefit us if we stop, recognize it and cooperate with it. There is a cyclical pattern of sound reception and expression between the voice, the ear, and the brain. Breath is important for us to live. Our breath stream provides one of the rhythmical systems of the body and can support us to produce vocal sound. Breath is simply vibration in motion. We need our breath to support the energy of voice-produced sound, but not always cellular vibrational energy. We know the cellular energy lasts beyond when our breath stops. The Voice-Ear-Brain Connection incorporates the energy of sound vibration beyond vocal sound. It provides sound, thereby energy, to the whole body.

Sound Notes
Chapter Three, Summary Statements

- The Voice-Ear-Brain Connection is a newly identified system of the body.

- Life as manifested, is vibration. Vibration is motion; therefore, all life is in motion. Harmony of vibration unites life forms.

- The Voice-Ear-Brain Connection as a system brings the body to homeostasis or a place of equilibrium. The body is constantly working to restore itself to a natural balance, to adjust the body energies so that the systems are harmoniously working together. When one part of the system is out of balance the disruption is felt by the other parts of the system.

- Each cell has a resonance that needs to be in harmony with its surrounding cells.

- The body hears the sound of the cells with quantum vibrations.

- Einstein said that ether was necessary for the laws of physics to exist. This etheric level is important for understanding that an energy exists that helps connect everything in the universe. The Voice-Ear-Brain Connection exists at this etheric level supporting the physical body.

- Even though energy of a movement may not be audible to the human ear sound is still emitted by the waveform.

- With sound work it is important to entrain the weaker frequencies to the stronger responses in order to balance the responses.

- There are five levels of human frequency spectrum: physical, etheric, astral, mental, and causal.

- Our physical senses and our consciousness make us think the world is as it appears. The true nature of reality is beyond our ordinary sensory capabilities.

- The Voice-Ear-Brain Connection is an interface of energetic exchange between the physical body and the vibrational sound energy within the surrounding environment.

CHAPTER 4
Sound and the Body

*T*he cellular level's response to sound energy is very important to the connection between the voice, the ear, and the brain. The cellular level vibrations generated from this connection are vital to our everyday functioning. Although each small part within the system should and can be studied independently, the integration of all the parts is more important to the overall system and therefore to the whole person.

Consider the story of a boy who wants to ride a bike. There are two paths available to him. The first is to evaluate the bike. He needs to take it apart, see how the parts work together, make sure the spokes are good and the screws are tightly screwed in, and then determine the best way for him to sit. Next he needs to decide if the bike should have wide or narrow tires, and should it be a three speed, ten speed or more? These are all questions he could ask but all he really wants to do is learn how to ride the bike. With this evaluation he never gets on the bike. The second option is to get on the bike and try to move the pedals. If he practices enough he will learn how to stay balanced and roll forward. He will learn how to ride the bike. If he chooses the first approach it will take a long time before he gets to ride the bike. The second approach gets him on it quickly and accomplishes his goal. Sometimes, by needing to get all the answers and details using scientific exploration, the ability of our body to change naturally can be delayed. Our body is designed to heal itself and maintain that health. By exploring the pieces of the Voice-Ear-Brain Connection we can learn a lot of information, but when we go ahead and balance this system the body feels better very quickly.

There was an instance when my son-in-law, an emergency room physician brought home the swine flu to my daughter and grandson. I took a voiceprint and found each of them had some degree of swine flu. My grandson didn't appear to have a severe bout of the flu, probably because my daughter was nursing him. I put together some frequency combinations in order to help him. My daughter was nursing so frequencies were not advisable. And for my son-in-law, I put together a sequence of sounds designed to help move the virus out of his body more quickly, as well as some stabilizing sounds to help him feel better. All sounds require tone trials where the sounds are introduced to determine the body's response. Both my grandson and son-in-law appeared to do well with the trials. However, my son-in-law said he really felt the sequence of sounds more powerfully than the stabilizing sounds. He began using the sequence of sounds once an hour for four hours. In his words, "The sounds wiped me out, and I had to go to sleep". The next day he alternated between using the sequence of sounds and the stabilizing set. The third day I took his follow-up voiceprint and found he no longer had any swine flu activators. The virus had moved through him very quickly. My grandson maintained the swine flu activator frequencies through the following week, but had a weaker case of it. My daughter however, also maintained the activator frequencies, but she used no sounds and didn't seem to regain a sense of wellness for almost two weeks. My son-in-law was able to move out the virus more quickly by using the powerful sequence of sounds, thus regaining his homeostasis faster.

To feel in balance we need to resonate in equal measure with our internal self and our external world. Our bodies as well as the outside world are in constant movement. All objects we see are actually wave patterns formed into a relative permanence. These wave patterns are formed by frequency combinations. All cells respond to their optimal frequency or vibration. These vibrations correspond to sound. Whereas our body parts have their own sonic oscillations, so too can our thought processes be considered patterns of energy that have vibrational responses. Both our physical body vibrations and thought vibrations can be expressed to the environment surrounding us, as well as receive vibrational information from our environment.

What is Sound?

Sound has many meanings:[1]
- The sensation of hearing;
- Sensation due to stimulation of the auditory nerves and auditory centers of the brain, usually by vibrations transmitted in a material medium — commonly air — affecting the organ of hearing;
- Waves propagated by progressive longitudinal vibratory disturbances;
- Free from flaw, defect or decay; and
- Healthy, not weak or diseased — of mind and body.

If we combine a few of these definitions with our own, then sound is vibrational energy that impacts our mind and body to keep us healthy.

Sound travels in waves within some form of medium like air, causing the new surfaces they encounter to also move in waves. The particles within the wave move against new particles, causing them to vibrate. Within this movement the particles will sometimes crowd together until the new particles begin to move; then they move back or spread apart. Technically, this is called compression and rarefaction. The actual traveling wave of sound is created from the disturbance of the medium and not the particles themselves whose only job is to vibrate. Sound travels in longitudinal waves.

In the previous chapter I talked about the subjective responses to the acoustics of a sound: loudness, pitch, timbre, and richness. Sensation of the sound is also important. The basic tone that identifies the pitch of a sound is called the *fundamental tone*. The higher frequency portions of that fundamental tone are called *overtones*. Overtone frequencies which are whole multiples of the fundamental frequency are called *harmonics*. Harmonics help determine timbre and reinforce the sensation of pitch. The fundamental frequency of the initial waveform is the fundamental frequency of the final tone made and therefore defines its pitch. The human voice can control the overtones, but has little control over the fundamental frequency of the voice because of the body specific acoustical resonators within the body.

When discussing sensation of sound which includes the phase component, a phase shift is important to understand. Can the phase of one sound affect the phase of another? Cycles of waves have peaks and troughs, highs and lows. Two sound waves of the same frequency are said to be in phase if their peaks and troughs occur at the same time and place. They are considered out of phase if the peaks of one coincide with the troughs of the other. Sounds that are in phase reinforce each other, while those out of phase cancel each other. This will be important to us when we consider how we can tonally cancel out a sound we hear and/or produce.

The Effects of Sound on the Body

Much has been published on the effects of music on the body impacting heart rate, circulation, and other body processes. A large industry creating CD's based upon changing our body's responses has developed. The value of these musical effects is recognized in many different areas: mental health, psychology, wellness, and occupational therapy to name a few. Hospitals now have sound and music programs to enhance your surgery or overall hospital stay.

We receive external sound waves through our body. Our cells create sympathetic vibrations to help support a sound body. The vibrations create change physically, emotionally, mentally, and spiritually. While the voice reflects what these vibrations are doing to our cell structure the connection with the voice, the ear, and the brain demonstrates an even greater support system for the body.

Rhythm, the Seed, and Energy Movement

Although music is often thought of when you hear the word sound it is not music alone that is important within the context of the human body. Sound can be a single tone, or a frequency that is unheard. Sound can be a beautiful symphony that moves you beyond words, and sound can be a harsh high-pitched cacophony of frequencies that makes your skin crawl. Sound is all around us, yet often goes unnoticed. Sound can also be felt as a response to a vibration. The Seed of the Tree relates to our body rhythms. Within the cadence of the body there can be

86

harmony, melody, tonal patterns, and rhythmical patterns. The rhythmical activity of the external environment can also influence our body rhythms, for example in kinetic movements. The movement can be within the cell structure, a pulsing within the circulatory system, a beating heart, or a swaying of the physical body. To be in rhythmical harmony, these patterns need to support each other and support the energy of those patterns around you. There must be an alignment between the person's internal world and the world around them. Here is a quote that says it best, "Sound, music, the rhythm of speech, and the repetition of auditory patterns set the foundation of our understanding of the world around us".[2]

Yet, within the world surrounding us are other sounds. The earth, sun, and other planets make sounds that have been recorded with special instruments; the universe is playing its own song! Rhythms that include humming, sighing, crackling, ticking, hissing, or drum-like sounds have been reported. The sun's energy produces eighty harmonies measured in cycles between two to eight minutes. Here on earth, earthquakes record shock waves that show up as rings, and often a ringing sound can be heard. We have so many different rhythms, frequencies, and sounds surrounding us and impacting us daily.[3]

I love how Don Campbell in his book, *The Harmony of Health* discusses how as a child his body responded to every sound and music danced in every cell. He describes his body as an ear and that he was perplexed if those around him weren't moving as he was.[4] He began to see at an early age that the harmony of the world was within the sounds that you heard and moved to. That body movement combined with what he heard made a connection for him. Demands of society may alter these early-learned skills, but they nevertheless remain very important to our whole being.

The rhythms are contained not only within us, but all of nature is rhythm. There are rhythms of the universe, for example, we travel around the sun and our moon travels around the Earth. The atoms and smaller particles of the universe are in constant movement of varying rhythms. They also are of varying lengths but all are based upon movement. We have a life cycle, birth to death. We have tissues and enzymes that are replaced frequently, but in patterns. We have organ cycles: heart beats, ovary cycles, breath streams, and brainwaves. And each cell has a pattern that defines it. Then for every frequency produced, there are harmonics and sub-harmonics for that sound. Our rhythms and frequencies are very complex.

The Voice-Ear-Brain Connection incorporates rhythm in many ways: cellular rhythms, patterns of frequencies received by the ear for movement and hearing, patterns of sounds emitted by the voice, and patterns used by the brain to support the body with change for movement, sound output, and cellular repatterning.

How does the brain support the body with change from sound vibration? Oscillations of the brain's direct current field or brainwaves help direct and support body change. These patterns begin in the thalamus. They travel through and spread out from the brain through the circulatory system and along the perineural cells encasing the peripheral nerves. In this way, the vibrations reach all parts of the body. The perineural cell system is also a direct current system. The perineural cell system supports all of the connective tissue networks. Brain waves can be measured because of the rhythm and synchronicity of the direct current vibrations through the neurons of the brain. Brainwaves can also cause the fields around the neurons to change rhythmically. In this way, brainwaves regulate the nervous system and states of consciousness.[5]

The physics principle of entrainment demonstrates that when two rhythms with nearly the same frequency are connected, both evolve into one rhythm. Because brainwaves are not constant in frequency or rhythm patterns there are times during the down time of the patterns that free radical frequencies or rhythm patterns can sneak in to alter the rhythm, creating an entrainment to the old or newer rhythm. The newly evolved pattern is then sent throughout the body. Humans subconsciously adjust our modes of communication by our words, body movements, or visual choices, and by reacting to the sensory style of those around us. We also subconsciously adjust to our newly evoked cellular rhythms. By using The Davis Model of Sound Intervention, the body searches to balance itself by using the concept of entrainment. By merging a less energetic pulse with a more energetic pulse our body then searches for a more efficient state from which to function. The rhythms of the cells are trying to communicate more efficiently.

This personal entrainment can be used to support those with energy imbalances such as attention deficit hyperactivity disorder (ADHD) or autism spectrum disorders. The Davis Model of Sound Intervention works to repattern the out of sync energy by entraining the person's energy to a more functional pattern based on those around them. If left by themselves those with attention deficit or autism function quite well within their own energy patterns, but not necessarily in the energy patterns as defined by the social situations demanded of them. Society imposes a different set of requirements for movement and energy responses so that individuals with these disorders stand out.

We had an interesting case of receiving out of sync energy recently in our office. One of our very sensitive adults who had a traumatic brain injury began to enjoy a quality of life that she hadn't had since her accident a few years ago. The process was long for her but we got her to a stable position. She recently emailed me and related the following: "This very strange phenomenon has been happening. We have recently hired a young woman to help us around the house. She has a pronounced stutter. She doesn't stammer, but she will repeat the first syllable of certain words two, three or four times before being able to say the word. I noticed this after spending some time with her. Within a few days, I found myself stuttering, saying words wrong, not being able to get the words out, and getting brain confusion; basically a lot of the symptoms I had before. The problems seem to slowly abate after she leaves for the day but resume when she comes back the next morning. They are not abating completely each night. What do you make of this, and is this something that is temporary or really going to hurt me?" My response to her was this: "You are a very sensitive person as we know. The physics principle is entrainment where the source of greater energy entrains the weaker source. In this case, your worker is the stronger one and is entraining you to work at her pace. Her pace is off for you and since you are finding it harder to adjust after she leaves, she in effect is creating a reversion of what we did with your Listening Training Program. This will not be temporary if you continue to be around her."

I copied the other therapists in my office about this unique case and the next day, my head sound therapist came to me and said, "Dorinne, I found your response to our client really fascinating. Let me tell you what has been happening to me. We recently had a stuttering client in and both he and his parents are very pleased with the change. However, when he comes in for therapy, I find myself beginning to stutter myself for at least one week after he leaves. It will clear up until he returns. Thanks for making me understand why." This therapist then went on to say that this stuttering client wasn't always maintaining the changes and we were questioning why. His father also is a severe stutterer. When the child is home without therapy for a while and is around his father and hears him stutter, our hypothesis is that this same phenomenon as with my other client may be occurring. The greater energy of his father may be entraining his weaker, not yet fully supported on his own energy back to the stuttering mode. The rhythms and energy movement within the entrainment that occurred was powerful for these people.

The rhythms of the body are often noticed through pulsations of our heart, breathing, etc. Whenever one of these pulsations is disturbed our whole body is disturbed, and illness or disease can result. We depend upon the regularity of these pulsations and they can be disturbed by shock, fear, medication, external events and more. The rhythm is broken and we work hard to regain the rhythm. In some cases, the rhythm is never reclaimed. Sometimes disorders remain as a result of these broken rhythms and become manifested as the symptoms that are associated with disease.

In a recent study at Stanford University, researchers suggested that brain cells must follow specific rhythms for proper brain functioning and when the rhythms are not correct, certain diseases like autism and schizophrenia can result.[5] The results demonstrated that by tuning the vibrating frequency of particular neurons the ability of the brain to process information was affected. The brain cells produce the frequencies that control the behavior of other cells that use the original rhythm to stay in balance. If the original frequency is disturbed the subsequent frequency will also be impaired. The conclusion suggests that the frequencies found for individuals on the autistic spectrum appeared to be at the wrong intensity, typically too much, while those with schizophrenia had too few neurons stimulated with the frequencies.

The rhythms of the cells are key to beginning the process of entrainment as well as being able to feel balanced. Rhythm stability is the Seed of The Tree of Sound Enhancement Therapy because at any time in the therapy continuum one can return to stabilizing the body rhythms.

In the last chapter, we looked at the five elements of energy movement. Ether is the element where sound is key and the voice produces the sound. However, when considering the rhythms of the body to the elements the relationships are:[6]

Earth: Very Slow to Slow
Water: Moderately Slow to Moderate
Fire: Moderately Fast-to-Fast
Air: Very Fast to Erratic
Ether: Silence to Space

The rhythm of ether is silence which is the place we hear our inner sound. The silence supports finding that sound which allows us to connect with the universe (or space) by using our voice to Ototone what the body needs which is provided by the ear. It is so exciting to see how all the pieces that have been there for hundreds of years all come together with the Voice-Ear-Brain Connection.

Natural Energy Phenomenon

Within nature we have many natural phenomena. Rhythm for example, is a hidden law of nature. "There is no movement which has no sound, and there is no sound which has no rhythm."[7] Rhythm, movement, and sound all work in harmony. We already know of the rhythm of the body, but behind any form that we see is a rhythm of the movement of the energy source. Harmony is a function of the right rhythm; disharmony is a manifestation of a disorder within a rhythm. A person's rhythm can attract us to another person or make us shy away and avoid that person. We feel comfortable with someone when his or her rhythms match ours.

Some people say we must follow our own rhythms. I have a funny story that anyone who travels with me understands. *The line forms behind me.* Whenever I go somewhere I must keep my own rhythm or pace. When I do I am always first or second on a line. It is now a joke that as soon as I get on a line, if we turn around there are at least five or more people behind me. If someone interrupts my rhythm by taking his or her time getting out of the car for example, then I will be at the end of a long line. This rhythm has been in existence my whole life. When I try to adjust to someone else's rhythm, mine is invariably distorted.

How we respond to the rhythms and timing of sound perceptions is important to us developmentally as well as for communication with others. For example, Timothy Roberts, PhD, from The Children's Hospital of Philadelphia suggests that children with autism detect vowel sounds and tones a fraction of a second more slowly than neuro-typical children. He hypothesized that speech and other sounds are processed too quickly for children on the autism spectrum which would impair their communication skills. Upon testing he found a delay of twenty milliseconds (1/50 of a second) for those on the autism spectrum. When mismatched sounds were then introduced the delay ranged up to fifty milliseconds. He suggests this delay in auditory processing may lead to a cascade of delay and overload of processing sound and speech.[8] There is a breakdown within the Voice-Ear-Brain Connection, which is demonstrated by this measured delay. Although not researched yet, many parents of children on the autism spectrum and other learning challenges have commented that their child's response time has been increased after using a sound-based therapy.

Dr. Gerber in his book *Vibrational Medicine* discusses the details of resonance at the atomic level:

> "Resonance is another phenomenon which occurs through-out nature. At the level of the atom, we know that electrons whirl about the nucleus in certain energetically defined orbits. In order to move an electron from a lower to a higher orbit, a quantum of energy with very special frequency characteristics is required. An electron will only accept the energy of the appropriate frequency to move from one

energy level to another. If the electron falls from the higher to the lower orbit, it will radiate energy of that very same frequency. This required atomic frequency is referred to as the 'resonant frequency'. ...Atoms and molecules have special resonant frequencies that will only be excited by energies of very precise vibratory characteristics... The vibrational mode of the physical body is a reflection of the dominant frequency at which it resonates."[9]

The words photon and phonon have been used when discussing energy properties and are natural phenomena. Photon refers to the smallest unit of energy of a light vibration, and phonon refers to the smallest unit of energy of a sound vibration. For The Cycle of Sound we are more interested in phonons than photons. However, it is interesting to note that the movement of sound through materials can alter the optical properties.[10] It is not uncommon for clients to report a change in visual responses after listening to sound.

Sound is thought of as wave disturbances passing through a medium (solid, gas, liquid). But in quantum theory we should also look at phonons as particles within the wave-like disturbance. They have the ability to collide with one another and hold onto other energy particles like toxins. In this way, heat can also be transferred with phonons. Phonons communicate with other phonons and are powerful units.

Another phenomenon of nature is coherence: parts in an orderly relationship or balance. All of nature wants to be orderly. When order is disturbed the surrounding natural pieces become unbalanced or out of order. The phonon energy becomes unbalanced. Our many body systems will be disturbed resulting in our Voice-Ear-Brain Connection being out of balance as well.

We measure the voice for BioAcoustics by looking at vocal coherence. I watch for coherence in spontaneous otoacoustic emissions and in my auditory testing. We want there to be order to the sounds we receive and the sounds we emit. "Coherent structures give rise to coherent energy fields which feedback on structures to stabilize their coherence."[11]   There is order in our cell membranes, muscles, connective tissue, brain waves, and heartbeats; our entire body should have order or coherence. Coherence links our body's subtle rhythms, as well as helps radiate energy outward from the body.   Coherent signals are resistant to interference from surrounding random fields of energy.   Being coherent demonstrates stability because the feedback loop is operating as it should.   Every molecule in the body can act as both a source and conductor of sound that spreads throughout the body, so consider the body as the original 'surround sound' device. Coherent systems lead us to feeling balanced, whole, and healthy.

There are so many natural occurrences and phenomena that are within and without our bodies.   To find the solutions to where our challenges lie we need to look in both areas.   But wherever we look we know we need to be resonating at frequencies that demonstrate stability as coherence.   We should be looking at The Cycle of Sound as an important system of coherence for our body.   The give and take of sound frequencies keeps us feeling balanced.   The ear picks up and sends out sounds.   The voice demonstrates the irregular patterns and the brain can send out the correcting patterns.   The Cycle of Sound is the Voice-Ear-Brain Connection!

Cellular Checks and Balances

We now know that everything in the universe is in motion; all material is in motion.   In quantum physics, all matter has movement because the matter particles behave as waves which are vibrations that repeat themselves.   If everything vibrates then a frequency is emitted which is sound.   Humans typically think of sound as the frequencies picked up by the ear.   But if everything around you is a frequency or sound, then why are you not hearing it? Because it is necessary to extend beyond the physicality of the ear and think of vibration or waves sent through the body to the brain.   The brain recognizes the frequency impulses outside of the frequencies of the ear.   The body, in the give and take of the frequency cycles recognizes the vibrations as quantum movements and searches for ways to create harmony.

Kevin Todeschi in his book *Edgar Cayce on Vibrations: Spirit in Motion* sheds light on this idea through sharing a quote by Plato from *The Republic*:

> "'A world in which men are imprisoned by chains attached to their legs and necks – unable to move or to see anything but the movement of shadows cast on the wall of the cave before their eyes.' In this world of chains the prisoners only see the reflections cast by the shadows of reality behind them. This concept of only being able to see a shadow of true reality is relevant to the topic of vibrations in that our senses provide us with a perception of the material world that is essentially a shadow of the truth. We think we perceive reality when in reality, we perceive the effects of the vibrations we encounter rather than being able to perceive those vibrations directly."[12]

We also think we are aware of the reality of sound because we hear with our ears, but actually we perceive the effects of the sound vibrations all around us through our entire being, through our received quantum perceptions of the vibrations.

In 1932, Edgar Cayce shared that what we experience in the physical body as our senses is simply an alteration of vibration attuned to those perceptions within our consciousness of the physical body. He said we have one force which he called 'spirit' that connects us to the material world. At its foundation is movement and vibration. The movement of these forces, or spirit, brings about harmony of the senses, which in turn affects the physical body.[13] His readings suggest that "one primary purpose of life is to evolve in consciousness in order to be more in attunement with the vibration of the one force that moved all of creation into being in the first place." [14]

Everything in the body has its own frequency: every organ, muscle, tissue, and cell. Every molecule and atom also has a frequency, giving out its own vibration that is distinctly unique and important for its own function. Ideally the vibrations should harmoniously connect with other systems in the body. When one of the body parts is not working properly, the emitting vibration is heard as discordant to the other body parts and impacts the entire physical body. In order to self-heal, the body searches for a way to tune up the discordance so that the body can return to harmony or optimum vibration.

Consider how difficult this is for our bodies! We have nearly one hundred trillion cells in our body. Within each one of us we have more cells that the entire population of the earth by 1,500 times.[15] Cells are part of action groups which are tissues grouped by their shape, size, job, and rate of cell division. Cells are constantly working to renew and repair our body. Their job is to keep us healthy and functioning well. A harmonious relationship with the cells around each other makes the job easier.

Many consider this idea to be relatively new. However, Herbert Frolich in 1978 explained that "If... a cell shifts its frequency, entraining signals from neighboring cells will tend to reinstall the correct frequency. However, if a sufficient number of cells get out-of-step, the strength of the system's collective vibrations can decrease to the point where stability is lost. Loss of coherence can lead to disease or disorder."[16]

When the body is discordant or out of balance the atomic forces within the cells are not working in cooperation. As a result, the body reacts and symptoms occur like headaches, fatigue, or inflammation. The body is letting you know something is not in harmony and the body searches for a way to correct the discordancy. This has been the benefit of sound-based therapy. The irregular vibrational patterns are identified and the correct sound-based therapy supports the body towards self-correcting.

Candice Pert in her book *Molecules of Emotion: The Science Behind Mind-Body Medicine,* discusses that molecules have receptors on their surfaces made up of proteins that act like sensors searching for other chemicals to bind to them. She compared this process to the right key opening up a keyhole. The new substance transfers a message via its own molecular properties to the new receptor. This process 'might be two voices... striking the same note and producing a vibration that rings a doorbell to open the doorway to the cell.'[17] Once inside, a series of biochemical responses begin a series of activities which impact our whole body. One type of these receptors is called a peptide. Ms. Pert suggests that "peptides serve to weave the body's organs and systems into a single web that reacts to both internal and external environmental changes with complex, subtle orchestrated responses. Peptides are the sheet music containing the notes, phrases, and rhythms that allow the orchestra – your body – to play as an integrated entity. And the music that results is the tone or feeling that you experience subjectively as your emotions."[18] The cycle of the body, the cycle of the cells, and the cycle of the peptides all provide for the give and take of the complex energy exchanges within the body.

At the center of each atom is the proton. Inside each proton are huge amounts of energy, discharged in nanoseconds. One such emission is called a gluon. Gluons flash and disappear during brief moments in time, but are very important because they help hold universal substances together. Gluons spin and charge the proton but have no matter.[19] Some energy source is creating them. We each possess the ability to create this energy, which in turn means we can work towards self-healing.

Consider the body to be one of checks and balances. The body operates as a flowing system balancing what is coming in and going out. Our healthy body needs this flow to be operating smoothly. The body learns how to adapt to new situations. Consider if you were to injure your ankle. The previously learned responses as to how to walk do not work with the injury because pain is incurred. The entire body takes on the process of helping the ankle heal. Not only does the muscle need to heal, but the surrounding tissues need to support the healing process. The body must direct nutrients and biochemical needs to the entire area to support the healing of the ankle. But the wonder of the human body allows that no matter how off balance the body is, we have the capacity to return ourselves to a balanced position.

Why might healing or returning to balance not occur? The reason is because our energy is blocked, even when there is significant energy within the system. Energetic healers try many different ways to unblock this energy. The Davis Center uses sound-based therapy which works directly with the energy blockages. The Diagnostic Evaluation for Therapy Protocol was designed to determine if, when, how long, and in what order any or all of the many different sound-based therapies can be appropriately applied. A protocol of therapies is suggested as a result of this test battery and the person follows this protocol for maximum success. Through whichever methods are suggested from the test battery, the vibrational energies of the therapies go to the weakest or neediest areas of the body. Although the test battery helps determine the therapies and the programs that will aid the person to make changes, the energy blockages may be so deep that only the body knows how to reach those blockages once the therapies begin.

Sometimes blockages are a result of emotional trauma. The amygdala in the brain connects our emotional responses with the cognitive centers. During a severe emotional trauma this center shuts down and the person functions mostly through their emotions. Their emotions drive their behavior, or in some cases, the emotions are lacking. If continued over time, such as with traumatic brain injuries or even with some disorders such as autism, the emotional blockage prevents the self-healing. Remember, emotions can be held at the subconscious level also. It is not uncommon for clients to break down crying while receiving some of the therapies. In one instance, a father called me anxiously into the session because his son was crying. He thought his son was being hurt. My first question was, "How often has your son cried in the past?" He said he has never cried! My response was that we were finally breaking down the blockage and that his crying was positive.

Cell Communication

Each cell has a cell wall which has many receptors. Although the cell wall is smooth the receptors can be considered sticky, as though they are searching for other materials to bind to it so that the cell can use the substance. These receptors determine which substances are allowed into the cell, but sometimes because of an inharmonious balance the receptor assumes it needs something that isn't beneficial. The body has a natural tendency to heal itself. Hippocrates said that nature cures disease. Our cells try to heal themselves, but if we constantly bombard our bodies with unnecessary or unwanted stimuli the cell's process isn't allowed to function as it is meant to do.[20] The quote, "My cells are literally talking to each other, and my brain is in on the conversation!" [21] fits well here. The cells *are* talking. Sounds are being made. The brain is receiving the conversation and tries to make it logical; the brain is trying to sing its song. The neurons in the brain are trying to be in harmony, and this harmonic organization is more prevalent in healthy bodies versus diseased bodies.[22] The disease appears to disrupt the communication between the nervous/immune systems and the mind/body. Sometimes the body can't heal itself all by itself, and a sound-based therapy supports the body with a better message from which it can begin to self-heal.

Communication between and among our cells is simply through organized patterns of vibration. These patterns are in mathematical arrangements. Mark Rider in his book, *The Rhythmic Language of Health and Disease,* shares:

"The common element... is the fact that the vibrational patterns common to all life are but different voices, not speaking different languages, but merely operating at different frequencies that simply require tuning into by shifting to the different stations, as on a radio receiver... Every atom, molecule, and cell in our body speaks the same language,... and we just need to shift... to the different mental frequencies to perceive this elegant symphony." [23]

Cellular language can be viewed as our body's voice singing its own song. As a result, we are always producing our own Signature Symphony of Sound. All the instruments must be in tune and following the conductor accurately for the body to function in maximum wellness. If the instruments are not kept in tune, and the external environmental and emotional influences keep knocking these instruments to produce sounds that are out of tune, then our entire system will be out of tune. If we keep our instruments adjusted, our body will have its needed harmony to function at its best possible level.

Once thought of as controversial, Harold Saxon Burr, a professor at Yale School of Medicine, proposed that a disease would show up in the person's energy field before the symptoms of the pathology. He felt that if an energy field was disturbed and could be detected and reversed, then the pathology could be eliminated or prevented.

In 1983, physicist Joel Sternheimer in his report *The Music of the Elementary Particles,* showed that each atomic particle corresponds to a frequency which is inversely proportional to its mass, and that this music of these particles means that we are composed of musical frequencies.[24] He demonstrated that each molecule of the body corresponds to certain melodies and that certain sections of music correspond to the melody structure of certain molecules.

> "Woven through the music of Beethoven, for example, are the melodies of ACTH (kidney molecule), Anti-tripsine (lung molecule) and Cytochrome (liver molecule). Beethoven died from complications arising from these three internal organs...each molecule in our body can be reactivated through resonance if it "hears" its corresponding molecular melody... Towards the end of his life, when Beethoven's friends asked him how he was doing, he answered, "my doctor can't do anything anymore for me, but the music can"." [25]

Because there are so many melodies in the world there are many possibilities of ways to resonate our body molecules. People appear to be attracted to certain melodies over and over again because their bodies seem to need the melody or general frequencies of the notes within the melody. Perhaps they are trying to heal themselves at a subconscious level. The music just feels right! Their molecules are responding to the demand for needed change. The same happens with unwanted sound. We tend to shy away or purposely avoid melodies that irritate us.

An evolving science is BioAcoustics which analyzes the frequencies of the voice. Everything in the body has a Frequency Equivalent associated with it and when irregular patterns are found through vocal analysis the brain can tell the body how to support itself towards natural wellness. The Frequency Equivalents disturb the energy field and work towards reversing the pathology. This science was developed by Sharry Edwards and research is ongoing. The Davis Addendum to the Tomatis Effect and the concept of Ototoning initially evolved from this wonderful science.

How can all this happen? Albert Szent-Gyorgyi, a noted scientist who won the Nobel Peace Prize for his work with Vitamin C, in 1941 proposed that the proteins in our bodies are semi-conductors. As it turns out, all molecules in our body's web of cells are semi-conductors. Semi-conductors form the basis for modern electronics as well; however, the human body is more complex. Water and electromagnetic fields support our life function. Energy flows through the electromagnetic fields and water can form structures that transmit energy. The molecules do not need to touch. The energy is simply transferred through vibration.

In *Sound Bodies through Sound Therapy*, I discuss in great detail how the cell and nervous system operate to support the flow of vibration throughout the body.* Herbert Frolich, a physicist from Liverpool, England found that our cell network, or living matrix as it has been called can produce coherent oscillations. The entire living matrix creates these oscillations or vibrations that move around and within the body, as well as those that are sent out of the body or cell structure. Each part – cell, tissue, organ, and so on – has its own resonant frequency. Each of these frequencies or movements creates its own signature of movement. With sound, this has been called a Signature Sound[26] and with electromagnetic response it has been called a biomagnetic signature.[27] Scientists and therapists have used this knowledge to create many new therapies that directly influence how the body tries to heal itself. The energy or vibratory patterns are simply repatterned. We are in a state of health when the vibrations are in resonance or totally connected without disturbance.

One small disturbance at the cell level can impact our whole body because energy fields are unbounded. They can project to far reaching areas. A fundamental law of physics tells us that energy cannot be created or destroyed, only converted from one form to another. The energy wave's strength may decrease with distance, but there is no

---

* Refer to Chapters 1 and 2 of *Sound Bodies through Sound Therapy* by Dorinne S. Davis

known end to the energy field. So a tumor growing in one area of the body can have an impact on another area. Cutting the tumor out creates yet a different disturbance. I liken this to the story of The Princess and the Pea. The Princess could not sleep. She was on a bed of about thirty mattresses but was uncomfortable. As the story goes, a small pea was under the bottom mattress and the Princess could not sleep because she felt the pea. By feeling the pea, they knew she was royalty. In real life, she would be called hyper-vigilant. If however, we detect and reverse the tumor's vibrations as suggested by Harold Burr, we support overall wellness, overall resonance.

If the pea or over-sensitivity is removed we return to normalcy thereby maintaining resonance. Sound-based therapies can support more balanced sensitivity, restoring and maintaining one's resonance.

Candace Pert has said,

> "I believe that the receptors on our cells even vibrate in response to extracorporeal peptide reaching, a phenomenon that is analogous to the strings of a resting violin responding when another violin's strings are played. We call this 'emotional resonance', and it is a scientific fact that we can feel what others feel. The oneness of all life is based on this simple reality: our molecules of emotions are all vibrating together."[28]

And within these molecules, we can attend to what we need by listening to our silent selves. It's more than an inner voice telling us right from wrong; it is a sense of harmony, a sense of balance, a sense of feeling right with the world. In this way, our Signature Symphony of Sound is balanced and harmonizing with the world.

Cell Patterns and Responses

We have been discussing the frequency energy of the body and its parts. The concept of subtle energy has been identified within the field of homeopathy. Homeopathy uses remedies that are called subtle energy medicines because they contain the energetic frequency or vibration of the plant used in its preparation. Assuming that each plant has its own particular energy pattern or vibration, then homeopathy matches the frequency of the plant extract with the frequency of the illness in the person. Only the subtle energy of the proper frequency will shift the body towards a better state of health. In other words, the body is moved towards resonating in the correct frequency or vibration. For homeopathic remedies, it is the vibrational frequency of the plant and not what it is made of (molecular properties) that creates the healing benefits. By using sound frequencies one eliminates the concept of molecular properties except in certain circumstances, and focuses on the frequency vibration needed by the body.

Deepak Chopra in his book, *Quantum Healing: Exploring the Frontiers of Mind/Body Medicine*, discusses that a cell is a "memory that has built some matter around itself, forming a specific pattern".[29] The specific pattern of the cell is designed around the chemical reactions contained within each cell. So he talks about how the cells need to stay in tune. If one cell's tune is distorted, everything is thrown out of tune. The whole body will not function properly if one piece is out of harmony.[30]

This concept is similar to the one established by Dr. Guy Peter Manners who developed the field of Cymatics®. Building upon the work of Dr. Hans Jenny, Dr Manners used sound-form patterns: sound organized as matter when stimulated with an oscillator. Beautiful patterns are created when a pulse is emitted to a plate and a wave forms a shape. What appears to be a solid form on the plate is also a wave, but both are created and simultaneously organized around the pulse of the sound. With sound there is no solidarity, so the patterns formed appear solid but are really vibration. Our conscious being sees the form more easily than a vibration. Within us it is this pulse that is our fundamental organizing principle. This pulse provides the sound of our cells and I believe, the sound of the spontaneous otoacoustic emission.

Researchers have been finding that our cells can indeed respond to sound. From cancerous cells established in a culture, if one sound was introduced the cell could accommodate itself better and appeared to be able to balance itself and keep its structure longer.[31] In 2004, a researcher from UCLA also discovered that cells can emit sounds and he called this sonocytology.

Deena Zalkind Spear tunes into the silent sound of people and animals. She said that a sound from a toxic substance feels like inharmonious or jarring beats of vibration against each other. The energy field is disturbed by this ultimately resulting in damage to the body. When she transforms the vibrations into a harmonious relationship the chemical is no longer toxic or is less toxic.[32] She has moved the cell into harmony with the body.

This doesn't mean that the change seen is instantaneous. Sometimes the person feels worse initially. When dealing with the toxins mentioned above the physical toxins may come out through one of our major cleansing systems. If it comes out through the skin it may manifest as a rash or itch, or if through our elimination system as diarrhea or constipation. Previously suppressed emotions can also be released in excess such as the boy who cried during a listening session who never cried before. Sometimes the changes are great within the body, but only subtly expressed outside of the body, and the person simply feels better.

Sound, Music and Physical Issues

Edgar Cayce was asked why the higher notes in music impacted the emotions whereas the lower ones do not in a professional harpist. He responded that it was her own vibration or attunement and how she responded to them.[33] Dr. Alfred Tomatis emphasized the value of the higher frequencies for enhancing intellectual stimulation. He suggested that the ear supports cortical charge, or an enlivening of the brain.[34] To feel alive and mentally active one does best with higher frequency enhancement. The lower frequencies support our general body response and when exposed to low frequency sensation one feels more alive at that moment.

One reason young people today love listening to loud music with a bass boost is to feel great. They feel especially great while they are listening but when the music stops, after a period of time they need the music again to make them feel better. Once the listener is able to tune into the higher frequencies they stay more connected, more alive, more attentive, and more together as long as they stay connected with the higher frequency sound.

Cayce also expressed that the higher frequencies are healing frequencies when someone is troubled by emotional issues, especially when related to self-destruction.[35] For some people, or for particular disabilities such as autism, one needs to address the level of their body frequency response as a portion of their therapeutic process. Sometimes it is necessary to work towards a higher frequency response or change the reception patterns of the higher frequency response. Edgar Cayce discussed a device called a Radial Appliance which he thought equalized the vibrations in the body. People who could benefit from its use appeared to have a non-coordination of the correct vibrations needed for homeostasis.[36]

During another Cayce reading (440-6) the higher frequencies were mentioned:

> "As is known, the body in action – or a live body – emanates from the same vibrations to which it as a body is vibrating, both physical and spiritual. Just as there is an aura when a string of a musical instrument is vibrated – the tone is produced by the vibration. In the body the tone is given off rather in the higher vibration, or the color. Hence this is a condition that exists with each physical body."[37]

Some people need to visualize something such as a color for comprehension as their awareness may not be at a tonal or sounding level. In other words, the tone is a vibration and the color is a vibration, and both are at the higher frequencies. Both vibrations come from within, moving to the outward portion of the body.

Music and frequency have been linked with physical issues. Dr. Tomatis linked frequency with specific physical conditions. Basically the low frequencies affect the lower part of the body and the higher frequencies affect the upper part. Through the use of his bone conduction listening curve Dr. Tomatis linked specific ailments to the frequencies that are not in alignment within the curve. For example, ailments with the Head are related to irregularities at 6000 and 8000 Hz., the Tongue/Cervical Vertebrae at 4000 Hz., Shoulder Blade at 3000 Hz., Larynx, Plexus, and Dorsal Region at 2000 Hz., The Lungs and

Cervical Back at 1500 Hz., Heart at 1200 Hz., The Stomach and Medial Dorsal Region at 1000 Hz., Gall Bladder, Pancreas and Liver at 750 Hz., The Intestine, Elbow and Dorsal Lumbar Junction at 500 Hz., Intestine, Knee and Pelvis-Lumbar Region at 250 Hz., and Genitals, Pelvis and Feet at 125 Hz.[38]

Fabien Maman shares that he found several physiological correspondences with music. He found:[39]
- The musical note is linked with the acupuncture point;
- The melody acts at the molecular level;
- Harmony works with the ganglia and endocrine system;
- The resonance of overtones is linked with the nervous system, chakras and the subtle bodies; and
- Musical rhythm influences the circulation of blood.

This list is very interesting because of the processes involved within the Voice-Ear-Brain Connection. The musical note is generic or wider ranging than a specific frequency, and therefore makes sense that the action point is an acupuncture meridian point. The melody is the molecular level and within BioAcoustics, the science looking at the mathematical matrix of predictable frequency combinations, the practitioner must first look at the melody of the molecules. BioAcoustics then addresses the harmony of the body through the nerve cell ganglia and endocrine system. Additionally, the overtone resonance is important for the stability of the body. While BioAcoustics supports this overtone resonance, so too does Ototoning because the overtones spread vibrationally throughout the body will link all the subtle systems. The Voice-Ear-Brain Connection is the newly discovered subtle system of the body. Within this system, the musical rhythms influencing the blood circulation addresses the basal body rhythms which then fall into the Seed of The Tree of Sound Enhancement Therapy.

Higher Frequency Response

Sound travels through our nervous system. The glial and Schwann cell network surround most of the nerves of the body. When they wrap around the neurons they become white matter or myelin. They guide developing neurons to their proper places, help repair damaged neurons, and help regulate the ions and neurotransmitters. The vibration along these cell networks are with analog sound in direct current (DC) and not the digital sound of action potentials (AC). The brain uses analog sound for this nerve transmission. Analog transmission varies the voltage of the cell membrane in DC current potentials so that a shift in the information relayed is of a different type. Analog is a slower transmission system than digital transmission. The glial cells are found in the brain and spinal cord and the Schwann cells are in the periphery. The glial cells seem to function as an interface between the acupuncture meridians and the central nervous system, i.e. between the electro-interactive interface of the cell network and the nervous system. Perhaps the connections between the voice, the ear, and the brain work in a similar way.

In one of his readings, Edward Cayce related that each of our senses send information to the brain at a different vibration. The ability to produce speech, however, is the highest vibration. For example, the sense of taste sends information to the brain three million times less than what is necessary to produce hearing or sight, and speech is three times greater than hearing or sight.[40] It's interesting that he saw the voice having the highest vibration. The voice demonstrates itself as the body stabilizer. It provides us with a way to identify our imbalances and also provides us with a way to change our inner vibrational patterns. The vibrations within our body and those we give out create a cycle of demonstrating our weaknesses while supporting our wellness. Perhaps it is through vibration that we will have a better understanding of our bodies.

In the past it was thought that nerve transmission was simply sent via signal from one nerve cell to the next by a synapse. While this is a possibility, it is now known that the brain has some control over this ability. The nervous and immune systems can mimic each other. The brain must work together with the molecules and biological necessities like amino acids and peptides to make the process work. While the neurons travel along the nervous system, the neurotransmitters are touching every cell. These neurotransmitters provide information about emotions, memories, and desires. They support our immune system. These impulses can reach all cells in the body if necessary.

The body has the capability to mirror any emotional or mental event thereby affecting our vibrational patterns which eventually can lead to illness. The body listens to the mental, emotional, and physical message, and will react. The reaction triggers vibrational dyssynchrony within the body which triggers an imbalance in the Voice-Ear-Brain Connection. The natural cycle of sound is disrupted and the body reacts. The neurons and immune system share the same lines of communication and can break down when overloaded. This means that learning and memory can be affected when another system is overloaded.[41]

Perhaps we can see this best when a child is developing and has a gluten sensitivity. If the body is always protecting itself when overloaded on gluten the child's ability to learn is diminished. The body needs to divert itself from learning and focus on supporting the body with its gluten sensitivity. Once the gluten sensitivity is removed by eliminating it from the child's diet, the body then has a chance to work on learning and development.

One important piece of the Diagnostic Evaluation for Therapy Protocol (DETP) revealed through The Tree of Sound Enhancement Therapy is the important emphasis on wellness or the Head of the Tree, for this piece can determine if the body is unbalanced as evidenced with vocal analysis. Through the vibrational energy of the voice one is able to help determine and repair our irregular patterns.

Going a step further, if the voice has the highest vibration, what can thinking in words accomplish? Both Stanford University and Princeton University have researchers who suggest that thoughts affect subtle energy.[42] This mental response could trigger a dyssynchrony within the body. Could this highest vibration also affect our spontaneous otoacoustic emission? This would make for an interesting research project. How does the mental and also emotional response affect this Voice-Ear-Brain Connection? We don't know yet.

Within the highest form of mantra yoga a level of special attainment is reached when the mantra tone is an idea of sound in the mind and not vocally produced. This voice in the head has produced its own form of subtle energy. Our thoughts are energy fields. We have both our own thoughts as well as receive thoughts from others. How often have you heard someone say something that you were just thinking? You helped direct his thoughts into words. We can shape our reality by our thoughts. Our thoughts can influence our lives. Within this thought process the inner voice is part of our Voice-Ear-Brain Connection and can create vibrational dyssynchrony as well.

Sound Notes
Chapter Four, Summary Statements

- Our body is designed to heal itself and maintain that health.

- All objects as we see them are actually wave patterns formed into a relative permanence. These wave patterns are formed by frequency combinations.

- Sound is vibrational energy that impacts our mind and body to keep us healthy.

- The earth, the sun, and the other planets all make sounds that have been recorded with special instruments; the universe is playing its own song!

- By using The Davis Model of Sound Intervention the body searches to balance itself by the concept of entrainment.

- The rhythms of the cells are key to beginning the process of entrainment as well as being able to feel balanced.

- Harmony is a function of the right rhythm within the body; disharmony is a manifestation of a disorder within that rhythm.

- There is order in our cell membranes, muscles, connective tissue, brain waves, and heartbeats; our entire body should have order or coherence. Coherent systems lead us to feeling balanced, whole, and healthy.

- We think we perceive the reality of sound by our hearing the sound through our ear, but we perceive the effects of the sound vibrations all around us through our entire being, through our received quantum perceptions of the vibrations.

- When the body is discordant or out of balance the atomic forces within the cells are not working in cooperation. As a result, the body reacts and symptoms occur like headaches, fatigue, or inflammation.

- Sometimes the body can't heal itself all alone, and a sound-based therapy supports the body with a better message from which it can begin to self-heal.

- People appear to be attracted to certain melodies over and over again because their bodies seem to need the melody or general frequencies of the notes within the melody.

- Everything in the body has a Frequency Equivalent associated with it, and when irregular patterns are found through vocal analysis, the brain can tell the body how to support itself towards natural wellness.

- The voice demonstrates itself as the body stabilizer and provides us with a way to identify our imbalances, as well as provides us with a way to change our inner vibrational patterns.

- The nervous and immune systems can mimic each other. The brain must work together with the molecules and biological necessities like amino acids and peptides to make the process work.

- The voice demonstrates itself as the body stabilizer. It provides us with a way to identify our imbalances and also provides us with a way to change our inner vibrational patterns.

- Our thoughts are energy fields. We can shape our reality by our thoughts.

CHAPTER 5
The Voice

The voice is listed first in the Voice-Ear-Brain Connection because of its importance to the system. The voice reveals so much about the person, yet provides the mechanism for self-healing. The voice alone can be helpful to a healing process, but when sounded with the correct tones, the healing becomes more of a naturalized process or a movement towards wholeness.

How many people have paused to consider what the voice reveals? I found the following story helpful in explaining the complexities of the voice. It was written by autism advocate Dan Coulter in an article entitled *Listening to Yourself:**

"A few days ago, my wife pointed out an article about listening written by teacher Andy Dousis who noticed his fourth grade students excluding a classmate from their activities. This classmate had trouble making conversation, so he sometimes pushed or grabbed others. He had other challenges, too, and often sobbed in frustration. While the other students were initially patient with this child, they became less and less tolerant as the year progressed.

In looking at his own behavior, this teacher realized that the good example he'd set at the beginning of the school year had slipped away from him. In September, he had

* A portion of the article, 'Listening to Yourself' copyright Dan Coulter 2008, published at www.coultervideo.com, was provided courtesy of the author.

put considerable effort into integrating this difficult classmate into the class, and his students had responded. But as the year wore on and he'd gotten busier he'd become impatient and spoken sharply to correct the child's inappropriate behaviors. The students were simply picking up their cues from their teacher. A good person and a good teacher, all it took to start fixing his approach was to listen to himself and realize what he was doing. Things got better for the lonely student and everyone in the class benefited.

This story brought to mind a conversation I had with a mother of a grown son with Asperger Syndrome... The mother explained how no one had known about Asperger Syndrome when her son was younger. She now looked back sadly at the way she had initially reacted to her son's difficult behaviors. One day her four year old daughter, after continually hearing Mom speak sharply to her older brother looked up at her mother and said, "If you'll be nice to Jim, I'll be nice to you." In that moment, her world changed. Even before a diagnosis helped her better understand her son's condition, her daughter helped her listen to herself and be more of the mother her son needed.

This mother wasn't alone. When my kids were little, my wife pointed out to me that I spoke to our son with Asperger Syndrome in a very different, and less patient, tone than I used with our daughter. I confirmed this listening to myself on some home movies. It's easy to respond with the first thing that comes to mind to fix an immediate problem, but in a way you might regret later. I learned to change my responses."[1]

The energy of the teacher's voice impacted not only the difficult student, but entrained the responses of the other students as well. Many people assume it was the language or words alone that made the situation between the young boy and his mother change. However, it was not just the mother's use of words but also the rhythm, pitch, inflection, tonality, and the meaning of those vocal characteristics within the words that had the most impact on her young son. And when the father said he listened to himself on home movies, it was more than the words he was using that grabbed his attention. He also noticed a difference in his tone which demonstrated how his voice reflected his whole response to his son, not just his words.

How do you handle difficult situations? How do you handle problems that occur at any moment? If you respond in a monotone and say, "I am upset" without any inflection, rhythm, or tonality, you are not representing being upset. Instead, it may seem more like you are lifeless. However, if you say "I AM UPSET!", one can feel that there is some impact there. Seeing this in print only minimally helps you understand. Hearing and feeling the sound response within the body provides a more powerful reaction. Our voice not only connects us to those around us, our voice also connects us to our inner core. It is our voice that lets us know we are feeling okay, that we are whole.

The Whole Person Response

Subtle energy systems work within the physical body but are sometimes viewed as the etheric body. The higher energy fields can connect to the physical cellular structure of the body through a network of energy flows or threads. Various vibrational characteristics flow into the body and influence the body at the cellular level. The chakras have been seen to support the flow of the higher energy fields into the physical body through the endocrine system. The Voice-Ear-Brain Connection appears to support the flow of higher energy fields into the physical body through many systems: circulatory, nervous, perineural (layers surrounding the nervous system) skeletal, and skin, because of the way sound energy vibration responds to the cyclical patterns of the give and take of the energy wave pattern.[2]

Deepak Chopra in his book, *Quantum Healing: Exploring the Frontiers of Mind/Body Medicine* discusses the importance of getting to the core of where healing begins, the deepest core of the mind/body system. He feels that consciousness starts to have an effect at the juncture of the mind and body. This juncture is deeper than the physical body.[3]

The Voice-Ear-Brain Connection incorporates the physical body, but also stimulates and incorporates the deeper emotional and mental processes. By stimulating the physical body with vibrational energy many changes can occur: better connections for learning, better physical responses for overall development, and better overall wellness. And, with this stimulation emotional blockages may also be released. The voice which reflects the change connects with the brain in order to reach the deeper consciousness. A total inner awareness of change engulfs the person to support the healing process. A better flow of energy is felt throughout the body. The person feels more together or balanced.

By using the Voice-Ear-Brain Connection the approach becomes a whole person approach. The balance of the energy signals within the body is more important than the physical symptoms. The physical symptoms do not manifest or show up until much later than the energy signal imbalances. Within quantum theory, vibration is central to how we function. We need to balance our body energy signals when considering a holistic approach to wellness.

Wayne Perry in his book, *Sound Medicine: The Complete Guide to Healing with the Human Voice* discusses 'the whole voice'. While he talks about this voice as being balanced and harmonious, he also says that "this fully realized voice emerges only when the inner and outer vibrational energies are developed and integrated."[4] This description is excellent. The voice is reflecting the balance between the inner and outer vibrations. The balance can only reflect how the body is actually functioning. The voice is only reflecting what the ear and brain are able to integrate. As Dr. Tomatis found, the voice can only produce what the ear hears. To find the final balancing piece, the ear emits the necessary sound as demonstrated in my work known as The Davis Addendum to The Tomatis Effect. Within the Voice-Ear-Brain Connection the voice reflects if the body is in balance. The ear tells you what you need in order to achieve balance.

Connecting to our Inner Voice

Mark Rider in his book, *The Rhythmic Language of Health and Disease* discusses how our voice is important to us. He suggests that,

"The human voice may... do for us what smell affords the mouse. When spectrally analyzed the same way amino acids and brainwaves are, the voice has been found to exhibit about four different patterns that are associated with health, mental illness, and physical illness. These patterns are no doubt recognized by us... for example, the soft, weak voice of some patients... with irritable bowel syndrome vs. the harsh, rapid voice of the hypertensive patient. Our voices then are... involved in bringing to the surface important features of what is happening on the inside."[5]

Fabien Maman in his book, *The Role of Music in the Twenty-first Century* offers the following:

"In the human voice there is an added element which cannot be found with any other instrument. The voice can be considered the premier instrument because its inflection carries not only the physical aspect (vocal chords, pitch of the note), and emotional colors, but also a finer, subtle element which comes from the conscious and unconscious will of the singer. The human voice carries its own spiritual resonance."[6]

And Khan in *The Music of Life: The Inner Nature and Effects of Sound* says, "the impression of the voice is more living, deeper, and has a greater effect than any other sound. Other sounds can be louder than the voice, but no sound can be more living."[7] The human voice has the capability of producing its own soul sound. This needed soul sound can be found from the emission of our ear.

James D'Angelo in his book, *The Healing Power of the Human Voice* says that the purpose of healing with the conscious human voice revolves around four specific sound types: natural sound, toning, chanting, and overtoning.[8] For hundreds of years the voice has been used as a healing instrument to help raise people to higher levels of consciousness. For some, the voice is a link between our conscious and subconscious.

We assume our decisions and our actions are made consciously. When I initially began putting together my notes of how sound-based therapies impact us, I was really floored to learn that the acoustic reflex muscle, with which I did my original work with Auditory Integration Training, begins to move before we begin to speak. In other words, apparently the subconsciousness of our body knows before our consciousness that we are going to speak. I'm finding this more and more as I consider the Voice-Ear-Brain Connection. Our subconscious is at the etheric level impacting the physical body which is the conscious level. We are aware in the subconscious level before bringing the process into our consciousness. This was demonstrated by scientists at the Max Planck Institute for Human Cognitive and Brain Sciences in Leipzig, Germany. They used a brain scanner to measure what happens just before a conscious decision is made. They found that many processes in the brain occur automatically without our consciousness getting involved. And they found that it was possible to predict the activation of brain signals seven seconds before the person consciously made the decision. The study demonstrated that decisions are prepared subconsciously before being acted on consciously.[9]

We use our voice every day in natural and expressive ways. Maybe you said, 'oww' when you stubbed your toe; or 'ohh' when you saw fireworks. Maybe you laughed when you heard something funny or screamed when things got too much for you. Making these sounds helped you express emotion. The 'ohh' for the fireworks enhanced the feeling of wonder. The 'oww' for the stubbed toe lessened your pain. The groaning, moaning, and grunting sounds that we make help rid our body of held energy and release it into the space around us. For example, when I am going through a difficult time in my life, I need to talk about it, and keep sharing it until I have released the energy. I have to do this in a positive way or all I do is rehash or recycle the negative thoughts. Others say that I often talk too much when something negative is happening to me, but for me, once I have released the negative energy and thoughts, I can move on in my life quickly – because I have used my voice to release it with positive intention. I have many female friends who are amazed at how quickly I can recover from some personal tragedy or trauma.

Many ancient cultures told how the universe was created by an all-encompassing sacred sound. The Bible says this sound was 'the Word'. The word *universe* also implies 'uni' as one and 'verse' as song or word. In order for the sound to occur a vibration or movement had to occur. The sub-atomic particles of the initial world had to vibrate so that the sound could occur. This 'Word' or one ultimate sound was the beginning of our connection to the universe as we know it. The vibrations began a cyclical pattern. Sound moved out from the source and the source absorbed the sound coming back into it. Very often these sounds were inaudible sounds but these vibrational waves became our early building blocks of our universe.

With the evolution of man, initial words were grunts, groans, hums, screeches, clicks and so on. Chanting was thought to precede speaking and toning to precede chanting. Human speech was thought to develop around 70,000 years ago and became necessary for our survival. Over three thousand languages have been created since then. It is the voice with its connections to the ear and brain that made that possible. The voice has always been available to us and is always there to demonstrate our soul. Our voice often sang to our body movements. Men might whistle while working, boatmen might sing while rowing, or women might hum while sewing. This was and is a communication with our inner nature and needs. Typically, because making extraneous vocal sound is thought to be different in today's world, we tend to be moving further away from this inner communication with our soul except perhaps when others are not around.

Carl Jung, the noted Swiss psychiatrist was interested in the inner voice. In therapy, he thought one's inner voice had the potential to tap into the unconscious thereby balancing both into a sense of wholeness. For Jung, the inner voice was necessary for the development of one's psyche. By balancing the voice, his patient was working towards becoming a more functioning person.[10] For some, our psyche is our soul.

119

The Voice and Body Vibration

Universally, humans function by beat patterns, i.e. heartbeats, cellular sound patterns, and breathing patterns. These initial patterns generate wave patterns to other parts of the body and external world. As the waves move out from the body, additional vibrations make the sound more complex in the form of overtones, harmonics, or undertones. These additional components arise out of the original fundamental sound. These additional qualities give each sound its wholeness or uniqueness.

These body patterns often have their own rhythm. Rhythm in music is simply a way of organizing sound. Rhythm is vibration or energy in time patterns. Our body functions in these time patterns. Without knowing it, we musically and physically experience this rhythm at our core.

The Seed therapies associated with The Tree of Sound Enhancement Therapy start with the wave patterns of the body. The vibrations released are expected to entrain the person's body rhythms or patterns to a wave pattern better suited for existence in today's world. For example, persons with ADHD may have body rhythms that function too quickly as compared to those around them, so that their hyperactivity is often seen as a disconnected response to everyday functioning. However, if they meet other individuals with the same rhythmical patterns of everyday functioning, they are not considered different. Groups of people are interconnected webs of vibrational energies. Some of the groups work harmoniously because their body vibrational patterns are in sync while others are not.

Deena Zalkind Spear says there is "no difference between physical and emotional vibration. It is all silent sound."[11] If, when working with a client, she doesn't feel the energy flowing in their acoustical field she considers the possibility of an emotional blockage. The emotion itself is not the problem, but holding onto the emotion creates a blockage in the person's energy field which overtime can manifest as disease in the body. This same disturbance can affect the people around them because "all is energy, energy expands, and energy affects energy".[12]

You can consider this in two ways: like attracts like and opposites attract. If like attracts like, the rhythmical patterns of the two entities work harmoniously and typically remain harmonious until some outside influence changes the patterns. If opposites attract, the over-activity of the one entity is calmed by the under-activity of the other entity. This may or may not bring about harmony. Sometimes, one can feel balanced by the other entity for a period of time, but then the relationship may break down quickly when an outside influence changes one of the patterns. The cycle of vibrational exchange between the entities is broken down and the relationship becomes disharmonious or dyssynchronous.

One way that humans share their vibrations with others is with their voice. Our voice reflects our whole persona. Olivea Dewhurst-Maddock in her book, *The Book of Sound Therapy: Heal Yourself with Music and Voice* says, "The voice reflects the mental, emotional and physical condition of a person: it is truly a parable of the soul. In the same way that the soul links the personality of the individual to the spiritual unity of the whole, the voice links the smallest wave or particle of energy to the energy of the universe."[13] So previously when my friend at dinner said that I was getting to the soul level, it is through not only the voice but also the connection with the ear as well as the Voice-Ear-Brain Connection that we reach the soul's essence.

Listening beyond the Words

Our voice is the body's own instrument and as such should be kept in tune. All musical instruments including the voice have three features responsible for sound production: a source of energy, a vibrational source that determines the sound and pitch, and resonators that supply the tonal qualities. The source of energy is our breath stream, the vibrational source is the vocal cords, and the resonators are the air cavities and structures within the throat, mouth, and nasal passages. To make a sound with these features we use three main processes: phonation, resonance and articulation. Phonation means to make the sound, resonance means demonstrating the harmonics of the sound, and articulation involves shaping and molding the vocal sounds. Most often we think of this shaping and molding as associated with speaking linguistic content, but articulation can also be associated with shaping and molding various other vocal sounds like humming or clicking.[14]

We share our emotions through our voice, but we also can identify irregularities within our physical body. Neither the words spoken nor the language uttered are the main issue when evaluating what vocal vibrations tell us about the person. The frequency, pitch and intensity of the voice often relays more information about the inner self than the words expressed. The words are simply the tools used to express what the person thinks should be related. How many times have you listened to someone sharing something with their words, but you were left feeling there was more to the story? You sensed that extra piece from their voice. So to truly listen to someone, you must learn to listen beyond the words.

The word *emotion* means a moving out from. We pick up the emotional components of a person's words by listening to the subtleties beyond their words. A human use natural sounds such as grunts, groans, and moans which provide an opportunity to really hear the person's voice. These are our natural sounds which our earliest ancestors used for survival and limited communication. Today we may be better able to hear what a person is saying through their laughter, crying, sighing, screaming, or whistling. Each of these carries and gives out the emotional component. Another natural sound of the human voice is humming. Humming is a way for the voice to express our internal vibrational energy patterns, and it is also a way for humans to put back into the body what the person needs to improve the emotional and physical body. This will be expanded upon once the process of Ototoning is introduced.

Singing is a natural human response taking us beyond speech. Singing does not have to include words. People often get too bogged down by the words of the songs and leave out the emotional pieces that provide more richness to the piece. By singing, the voice can produce more overtones to their voice than the words alone can. Babies often create beautiful musical patterns with their voices. Mothers often say that they understand their non-speaking baby by the sounds that they make. Babies freely give out their feelings by the sounds they make. Babies often use the 'mmm' sound for contentment. Mothers may use this sound to calm their baby when distressed. And my daughter, son, and grandson all made an 'mmm' sound while eating their food as babies. This meant they were happy with what they were eating.

As we age however, making weird sounds like humming while eating or clicking with your tongue while thinking is often frowned upon by society so these sound generated emotions and self-expressions are often repressed. Society forces people to use words to communicate. During a part of my diagnostic evaluation I ask my client to speak into a microphone. I tell them I want to capture their voice for a graphic printout of their vocal frequencies. Parents of non-verbal children are often concerned if their child does not use words for this vocal analysis. For the non-verbal child, being forced to produce words is not natural and therefore not their natural sound. This type of recording of their voice is not reflective of the inner core or self. If I can't get a child to hum or sing, I like to capture laughter or even crying. The sound is more reflective of their inner being, their true self.

Establishing our sense of self is guided by our culture or environment. In some societies, showing any emotion is frowned upon and the people's voices are often flat and limited in their frequency range. For many, the true expression of self is buried or repressed.

Recently, I reflected on what is important to me with music and singing. I love to sing; I grew up singing in choirs in church and school. I was going to be an actress and wanted to be in musicals. Do I have a favorite type of music? Not really. I see so many people today downloading selections of music to their iPods or phones and spending a lot of time learning the words of the songs. I've seen friends searching out the words to songs so that they can sing them correctly. Is this important to me? No, the words of the songs I'm attracted to are not what is important to me. I am more attracted to the rhythms, patterns, melodies, and tonal patterns of songs than I am to the words. It has never been about the words for me, it has always been about the sound.

One reason that toning, chanting, and overtoning have provided individuals with a sense of well-being are because the sounds produced are outside of a word context. The tones, pitch, inflection, intensity, and patterns of the sounds create the vibrational change needed for the person to feel differently. As individuals begin to use the sounds associated with toning, chanting and overtoning, the harmonics, tonal quality, frequencies, and resonances of the vowels and consonants begin to reactivate the person's energy. The voice becomes the resonating instrument for the body with the vibrations of the vowels and consonants. The voice creates the movement or vibration that moves throughout the physical body, stimulates the mind, and moves outward toward the surrounding environment. This sets the healing process in motion.

One research study revealed that "when one's personal notes are sounded, of all the sounds tested, those created by the naturally present overtones and harmonics of the human voice are the most supportive and beneficial to the body".[15] The results concluded that our voice produces our best healing sound.

Jill Purce, who teaches many workshops on sound and voice, discusses the connection between one's inner sounds and feelings. She has shared that if you liberate the voice you can liberate the body's tensions; and if you find your own note, you may find yourself.[16] Our voice is an extension of our inner self, which has been influenced by all the extraneous, external sound sources. Truly, we find ourselves within a cycle of sound integration.

Society is changing so quickly today. Technology has such a tremendous influence on how we function as a society. Preschoolers learn to use computers and electronic devices for communication so easily. Once they learn to read and write they can email, text, twitter, or do whatever else is the next trend. The only part that they need to know is the word by itself. The person receiving the contact has no clue as to the emotional or physical needs of the person writing. One reason the older generation (and I use that term loosely since some consider me to be within that generation) prefers to talk to people in person or on the phone is to hear the person's voice. We reveal so much with the voice that they understand more of the message with the vocal input. What will our future generations miss out on by not hearing the voice?

In addition to the four sound types for healing as suggested by James D'Angelo listed earlier in this chapter, a new concept is suggested. The human voice will not only reveal the emotional and physical needs of the person, but now the human voice will also be used to put back into the body what the body needs. This revolutionary concept will change the way people use sound as a healing modality for all future generations.

Vocalizing and Vocal Techniques

Some people consider singing to be a vocalizing technique. Darlene Koldenhaven in her book, *Tune Your Voice: Singing and Your Mind's Musical Ear* says:

> "Singing is a *learned* experience involving the proper ear to voice coordination to reproduce sounds and most importantly, the ability to AUDIATE (to think musically). *Pitch Matching* is essential in good singing and good musicianship. *Audiation* is at the core of accurate pitch matching. Adults and children who 'can't find their note' or 'keep the beat' would increase the enjoyment of their life considerably, by learning to audiate."[17]

This approach is meant for the singers of the world and differs from our ability to use our voice to self-heal because of the way Ms. Koldenhaven uses the word singing as a learned experience. When self-healing, using your voice should not be a learned experience. When Ototoning, one will have a good response if they can pitch match and be able to audiate, and would increase life enjoyment by reaching a level where they can produce their tone mentally, thereby supporting their body with the correct sound. However, our natural sound would be better than a taught or learned sound. Perhaps encouraging children when young to sing from the soul may enhance their audiation skills — training their brain to support what the ear and voice provide.

In today's world, you are judged by many factors when singing: breath control, tonal quality, nasality or lack of it, vibrato, etc. All these factors are typically taught by music instructors. Few singing instructors work on the natural expression of the voice, allowing the voice to resonate as the person wants to. We are also very concerned with learning the words of songs. Sometimes the melody, tonal patterns, pitch differences, intensities, and resonance of the voice or musical selection heard has more impact on us than the words. It is this ability to resonate within our body cavities first before expressing the sound to the world that is an important part of Ototoning.

Koldenhoven reports that the two most important things in singing are: breathing correctly and listening to yourself. She discusses listening to your mind's ear.[18] Each person should have the ability to hear something that they heard before and be able to imagine or create this sound in their mind's ear or mentally. All vocalizations are what are heard in the mind's ear. With Ototoning, this sound needs to be personalized to what the body says it needs.

Dr. Tomatis and the Voice*

Dr. Alfred Tomatis reported that the voice is a major contributor to the overall stability of the body through sound. He developed The Tomatis Method, which incorporates a heavy emphasis on the use of one's voice. First one learns to use their voice by listening through headphones while assuming a particular listening posture. Eventually, they learn to use their own voice to maintain the progress without headphones.

Dr. Tomatis's Listening Posture is important to the success of a person's listening in general, but is particularly important to the maintenance of the person's improved listening skills. This Listening Posture has the person sitting comfortably on a stool with their feet flat on the floor. Their back is straight and their shoulders are back or open. Their head is slightly forward so that their chin is tilted slightly downward and the apex of their head is the highest part of the body plane. One breathes deeply and exhales deeply. Then one begins using their voice. This voice is not a strain on the vocal chords but a free flowing sound on the breath stream. One should be able to feel the vibration of sound resonating through the spinal column. The larynx should be vibrating next to the spinal column. This should not be forceful, but in a relaxed

---

* The following section is summarized from the information in Chapter 8 in *Sound Bodies through Sound Therapy* by Dorinne S. Davis.

manner. With training one can produce many sounds on the breath stream without putting stress on the muscles in the oral motor area of the mouth. This activity is first done with the use of the Electronic Ear, the equipment used in the Tomatis Method (or with newly developed computer programs) to make change. Later, maintenance is accomplished through the person's own vocal practice. Using the voice with the correct listening posture helps bring about the cortical charge that Dr. Tomatis discussed frequently, thereby providing the body with a life enhancing energy. It is with the rhythm and pulse of the body that the Listening Posture helps sustain the body's sounds.

Results of The Tomatis Method and now other Listening Training Programs, have demonstrated that one can retain the ability to use high frequency information with their voice while listening through headphones. The use of headphones with music to support the voice is an external aid. This external aid creates the ability to bypass the negative, learned response. The last part of the Tomatis program trains the brain and the body to stabilize and control the voice and be able to support the person from within, thereby establishing his own internal aid.

Dr. Tomatis stated in his book *The Ear and Language*, "The human body is the instrument of language, and human language is the song that makes it resound. Man's body is the instrument man's thought uses to speak."[19] The expression need not be with words but can be expressed with the sensations of one's neurological system. The sensations are relayed to the listener who unconsciously uses his whole body to interpret and translate what was sensed. Dr. Tomatis calls this the song of our bodies. By allowing the body to sing, the song is transmitted to others through proprioceptive sensations. If the song is harmonious then the message is transmitted harmoniously; if it is discordant then the message is transmitted discordantly. The person's body image is formed by the sonic caresses of both themselves and others. The person's communication system is enhanced when they learn to play their harmonic keyboard effectively.

The song that Dr. Tomatis talked about was the cells of the body singing which has now been connected with the Frequency Equivalents identified through vocal analysis. Otoacoustic emissions were not identified until many years after Dr. Tomatis founded his method. The sonic caress of the body perhaps can be found with the expression and reception of the sound of the body, the sound felt within the body and the sound expressed outward towards others.

Techniques that Include Concepts of the Voice-Ear-Brain Connection

There are a number of different programs or techniques that include sound, music, and movement that have been helpful to many. Many similar yet different approaches have been suggested. Each may have one or two of the important concepts utilized within the Voice-Ear-Brain Connection.

As mentioned earlier, Dalcroze Eurhythmics is a philosophy and teaching technique which introduces musical art by linking the ear to the body and the mind. The method considers deep listening to be a multi-sensory experience moving the listener from music heard, to music felt, thought, and held as a healing force. Emile Jaques-Dalcroze encouraged students in the music field to feel the physicality of musical rhythms by engaging the entire body in vibration and structures.[20]

The method develops musical perception, increased awareness, improved attention, and greater control of musical expression. He had his students move their bodies to specific musical exercises. His method evolved to include "ways to harmonize the body's sensory systems, the emotions evocative influence, and the mind's memory and creative functions....When these three aspects of human behavior were all well exercised, people naturally came into balance".[21] They were unified in mind, body and spirit.

Eurhythmics teaches vibration to the whole body first and then moves to isolated movements. These isolated movements are first harmonized and then opposed, thereby creating a body symphony. The body needs to learn how to balance the opposing movements.[22]

Dalcroze felt that musical stimulation comes from two sources: the external world through audition, and the internal world through proprioception. Once both sources are stimulated, then perception, attention, memory, and action improve. These two sources are found within the ear itself. As has been demonstrated previously in other sections, the ear is more than a hearing mechanism. The vestibular section and semi-circular canals impact body movement, coordination, muscle tone, muscle planning, and proprioception. The cochlea responds to audition.

Deena Zalkind Spear developed the voiceovers or voice tuning technique. This is not a method that has been used by others. Ms. Spear uses her voice to make a change for people but she listens first to the person's own voice. She feels that each voice is a unique string orchestra.

> "Using that analogy, people who are stressed and ungrounded seem to have voices whose relative frequencies and harmonics are mostly in the violin range, with maybe a few violas tossed in for sanity. I discovered that by doing energy tunings, I could bring back missing frequency ranges as well as resonance into the speaking and singing voice, so it sounds to me like all the musicians in the orchestra have returned to their seats and begun to play."[23]

Ms. Spear also uses voice tuning when doing detox sessions.
> "Even though the silent healing usually succeeds in changing toxic substances into something more harmonious (benign), the voice can show me places where the body is still affected, and helps to guide me to areas of frequency that can use further fine-tuning. I don't necessarily need to know that a particular organ or other part of the body is in difficulty, because when I channel the energy to fix the acoustical disharmony, it repairs or contributes to the repair of the physical (or emotional) problem 'automagically'."[24]

As with the sound-based therapy associated with the Head of The Tree of Sound Enhancement Therapy, Ms. Spear does with her own voice what the science of BioAcoustics accomplishes with vocal analysis, yet at a more specific level of interpretation. Additionally, this same concept will be connected with both Ototoning itself and the Ototoning device of the future.

Jeffrey Thompson, founder of the Center for Neuroacoustic Research, developed a system called Bio-Tuning® and uses the sounds within your voice to support self-healing. He shares:

> "I knew I needed to use a person's own voice singing this fundamental note. This would release a unique set of harmonics and overtones, which only one's own vocal cords can produce – a voice vibration fingerprint. This voice vibration fingerprint is an exact pattern match of a person's essential vibratory template – the one used by a person's own "Biological Organic Intelligence", the Intelligence used to form one's body from two cells and to maintain it moment by moment thereafter. One also experiences a profound sense of subconscious recognition of the vibration frequencies of one's own voice.
>
> Using this voice-tone-frequency to the exact cent for balancing and healing, through a special Neuroacoustic Sound Therapy Table, one's body is kinetically resonated to the cellular level. The Sound Table has speakers built into it, so that one becomes one with the sound itself, as it were. Using this technique, it literally becomes difficult to distinguish where the body ends and the sound begins. There is a sense of melting into the vibration of one's own sound and one's own voice.
>
> To me, this is what the Mantra really was. A person would go to a great master who saw the entire world as vibratory patterns of energy and light. He was able to see the individual also as a unique vibratory pattern in the universe. The Master would sing the unique acoustic octave of this sound to the person, who would sing it back until known and memorized. The person would then meditate and chant this sound to him/herself—resonating from the inside out and balancing him/herself right down to the core of consciousness."[25]

Many additional techniques not associated with a specific person or program also contain one or two of the concepts of the Voice-Ear-Brain Connection. Vocal scanning is one such technique. The person initiates an 'oo' vowel sound. Then the voice slowly slides up and down as though sounding a siren. The body typically picks up a strong resonance point for one of the sounds made. This is a point where the body is in vibratory synchronization with the sound produced. Once this point is identified, the person holds onto the sound by continuously making the sound. Then, the sound is recorded for future reference and the sound is repeated if the sensation remains strong. When the appropriate sound has been found, the person's body feels the resonance. Some people have reported a tingling sensation.[26]

Toning the chakras is another method frequently written about and used by practitioners. Chakras are the vortices for energy swirling in our body between the base of the spine and the crown of the head. They are key points for the subtle energy movement within the body. When the flow of this subtle energy is disturbed, the person's energy is blocked or disrupted. Vocal toning helps restore the balance by using the resonance of the voice to alter the vibrational rates of the vortices. Tantra yoga or Kundalini yoga incorporates changing vowel frequencies. By exploring perceived vibrations while emitting a tone, each person can find the one tone which affects them the best or most, and they in turn tone that sound. When in a group the same tone is typically used for each chakra. Each person can change their tones intuitively as needed. No one set of tones or specific set of scales is established across the board for a better response.

Primordial sound meditation is a technique that originated in India. Primordial sounds are the unique sounds or vibrations of the universe formed into mantras based on the specific sound of the universe at the time and place of the person's birth. The mantra is repeated silently during the meditation thereby freeing the mind of external distractions in order to quiet the mind. This process brings out subtle feelings of conscious awareness and soothes the mind, body and soul. This technique has been promoted by Dr. Deepak Chopra.[27]

Throat singing has been practiced by Tibetan monks, Mongolians and Tuvans and produces two to four tones simultaneously. In order to do this, the Tuvans suggest listening for the overtones first and then shaping the mouth to create the sounds.

Overtone chanting had once been a secretive technique typically used in a religious context. By using the harmonic frequencies of our main vocal pitch the voice produces the sets of overtones simultaneously for a full spectrum sound. The sounds can be psalms, songs, hymns, or mantras. The effect can be powerful. They have been called 'the sounds within sound'.[28]

Resonance therapy is a psychotherapeutic approach that focuses attention on the healing capacity of the voice. At the heart of this therapy is the idea that we intuitively modulate our physical, emotional, and mental states by the sounds that we make. The voice accesses one's body energy. In other words, the voice gives off messages representative of the body's self. These sounds may be grunts, groans, hums, sighs, or exclamations. The tenets of this therapy are that sound affects the mind/body connection and that the voice jump-starts the person's healing process. As one produces the sounds, there is a feedback loop through the ear, through bone resonance, through cellular energy transmission, and through the body's processes. Using the voice as the body's modulator of energy, this intervention makes adjustments in the person's frequency tuning.

Voice Movement Therapy is promoted as an expressive arts therapy, but focuses mainly on vocal exploration. The approach incorporates singing, speaking, sounding, movement, breathwork, and personal identity. Ten vocal components combine to create the timbre of the voice. The objective of the therapy is an exploration of one's self and one's ability to communicate both verbally and non-verbally through the voice.[29]

Spin Off Methods

Many programs, techniques, and methods are promoted today as Tomatis-based programs. This term 'Tomatis-based' is over used and often used inappropriately. The details of the full Tomatis Method are proprietary to the method and are not revealed unless one gets certified in the method. However, people who have been certified have subsequently tried to copy the particulars of the method. For me, the important pieces are:

1) How the sound is introduced to the person: air conduction and bone conduction sound transmission through appropriate headphones;
2) The particular acoustic distortions in the presented sound: delays, precessions, and right ear advantage;
3) The length of the listening sessions—a minimum of 60 hours for brain change;
4) The gradual introduction of a high frequency sound emphasis through a special filtering process;
5) The correct Listening Posture; and
6) Integrating the active use of one's voice—using the person speaking with a feed-back loop through the equipment with the specific acoustic modifications.

Additionally, the program must provide the Tomatis Effect as mentioned in chapter two. A program can only be Tomatis-based if it incorporates all of these segments. To incorporate one or two of these requirements does not make a Tomatis-based program. Recorded CD's do not accomplish the effect of the Tomatis Method. It's like saying you can bake a cake with flour and water only when the recipe calls for flour, water, egg, baking soda, cinnamon and chopped nuts. Some programs incorporate pieces of the method but not sufficiently enough to make those programs fit into the Trunk of The Tree of Sound Enhancement Therapy. The Trunk's Listening Training Programs encompass all of the pieces of a Tomatis-based program. The use of the voice within what are called Tomatis-based programs will only be helpful if the other five components are present. Active voice work within a Listening Training Program will help the person maintain and support their overall changes over time. Only when the person begins to make the connections with the voice and the ear will the brain begin to know how to better support the learning and developmental pieces of The Davis Model of Sound Intervention.

The Science of Human BioAcoustics*

While the evolution of vocal sounding techniques and interventions introduced many ways of using the voice to support the body's needed energy, it wasn't until the science of Human BioAcoustics was developed by Sharry Edwards that the full impact of the connection between the voice, the ear, and the brain was possible. BioAcoustics as a science previously had been used with animals and more so with marine mammals. Around 1970, Ms. Edwards learned that she had a unique ability to hear the spontaneous otoacoustic emissions emanating from those around her. Later she also documented in three different sound labs that she had the unique ability to produce a pure sine wave with her voice. This is very unusual because humans typically produce overtones with our voice creating the individual characteristics that identify us.

With her unique skills, Ms. Edwards began studying the voice by frequency. She used what she heard emitted from the ear and connected the sounds to frequencies identified within the voice. She created a process whereby this information could be turned into a therapeutic application. By listening to specific frequencies one could repattern their natural form and function so that their body was better balanced. In other words, the frequencies listened to by the person were what the person's voice indicated were out of balance. The process works on re-establishing balance for an imbalanced system. She had connected spontaneous otoacoustic emissions with the irregular patterns of the voice and introduced a complementary sound protocol to balance the body.

To develop her process, Ms. Edwards studied the sounds of living systems. She learned that each living system emits an individual set of frequency patterns. She called this one's Signature Sound: a frequency representation of the whole body that is unique to the individual. In other words, everyone's Signature Sound represents their own body chemistry. One's Signature Sound is the body's vibrational fingerprint as it reflects the sound vibrational energy of one's body. The voice reveals the secrets of the body. Today this

---

* More information on BioAcoustics can be found in Chapter 9 of Sound Bodies through Sound Therapy by Dorinne S. Davis and at www.vocalprofiling.com.

process is beginning to be used by medical professionals to help identify wellness challenges.

The principle of BioAcoustics originated with the idea that the brain perceives sound and then generates impulse patterns known as brain wave frequencies. These frequencies are sent to the rest of the body through the neurological system. These neurological impulses help sustain the body's structural integrity and emotional equilibrium. When these impulses are disrupted the imbalance is manifested in disease or stress.

Human BioAcoustics uses the voice to identify the imbalances. After analysis, specific tones/frequencies are presented through analog sound presentation. The brain perceives these sounds, generates brain wave frequencies, and sends the balancing frequencies to the needed areas of the body.

Our body is like a blueprint comprised of the many frequencies needed to support itself. Ms. Edwards discovered that every muscle, compound, process, and structure in the body has what is called a Frequency Equivalent, which can be calculated mathematically. The body's ability to heal itself can be predicted mathematically within a matrix. This mathematical matrix displays the interaction between the body's systems and provides the specific frequencies that will support natural self-healing.

This wonderful science is used at the Head of The Tree of Sound Enhancement Therapy because it encapsulates the wellness and body support pieces needed within the Voice-Ear-Brain Connection. The Davis Addendum to the Tomatis Effect was only possible to identify once the technology of the science of Human BioAcoustics was invented. Ms. Edwards demonstrated the power of the voice as an expression of the imbalances of the body. Previously, researchers used the information from the voice to support emotional and to some degree, general physical issues. Otherwise the voice was used to make the body sing better and to restore a sense of harmony of well-being for the person.

With Human BioAcoustics the voice identifies not only the Signature Sound as discussed by Ms. Edwards but also the Signature Symphony of Sound as introduced within this book. Our body's instruments (cell structures) are playing their own symphony but the instruments are not always in tune. No matter how hard the conductor (the brain) of the symphony works at bringing the instruments into the corrected patterns, rhythms, or melodies with the correct modulations for emotional responses, if the instruments are not in tune the symphony will be discordant. If you have listened to a poorly tuned orchestra, you know that an orchestra can keep on playing even though not in harmony. The persons playing of their own symphony or Signature Symphony of Sound will reflect these discordances or body imbalances. Each of us continues to function even though we may be out of tune. The way the individual symphony is played may eventually lead to disease. Additionally, the external listener may perceive the symphony as being flat, sharp, or not in rhythm. The external listener may notice, 'You're out of sorts today', or 'You're not yourself today', 'Are you feeling alright?', or 'What's wrong? You seem down', or something similar. The voice provides those clues.

The Amazing Voice

The voice is our own instrument. It can be creative, harsh, melodious, expressive, and soothing. The power of the voice can express feelings, trigger emotions in others, or control others. The power can be with the tone of the voice or with the words that we use. We know that the voice represents our entire being, our Signature Symphony of Sound.

Through a combination of melody, rhythm, timbre, and other aspects such as frequency, duration, and pitch the human voice conveys who we are. The voice reveals one's energy. The sound generated is often a reflection of one's general state of being: our mood and feelings.

Our self-image, personalities, and emotions are reflected within our voice. Anger is typically relayed with a higher pitch, fear with irregularities in pitch, and sorrow is relayed with elongated vowel sounds. The voice responds instinctively to the energy and emotions of the whole body.

This powerful instrument, the voice, is our body stabilizer. It enhances overall body functioning. It provides a resonance through bone conduction vibration that allows the body to find ways to support itself. The body frequently does not know how to support itself without external help which may be in the form of sound based therapies, or with emotional, or psychological support. This book is introducing the concept of Ototoning which provides internal help by using the correct vibrational sound as introduced with the voice. Vocal stability is needed in order for the body to reach peak performance.

The voice reflects the imbalances of the body as evidenced by vocal analysis. The voice can also restore the body to a balanced position. Understanding the balance between what the voice expresses outwardly and how the voice can turn sound to good use internally, is important for understanding how the Voice-Ear-Brain Connection supports our body's wellness. The voice will only be successful in balancing our body if it provides the correct vibrational stimulation which is accomplished with bone conduction vibration, the correct posture, and knowing which sound is the correct sound to use as provided by the ear.

Sound Notes
Chapter Five Summary Statements

- The voice alone can be helpful to a healing process but when sounded with the correct tones, the healing becomes more of a naturalized process, a movement towards wholeness.

- The Voice-Ear-Brain Connection supports the flow of higher energy fields into the physical body through many systems: circulatory, nervous, perineural (layers surrounding the nervous system), skeletal, and skin.

- The human voice has the capability of producing our own soul sound. This needed soul sound can be found from the emission of our ear.

- Our body functions in time patterns. Without knowing it, we musically and physically experience this rhythm at our core.

- To truly listen to someone you must learn to listen beyond the words.

- As individuals begin to use the sounds associated with toning, chanting, and overtoning, the harmonics, tonal quality, frequencies, and resonances of the vowels and consonants begin to reactivate the person's energy.

- The human voice will not only reveal the emotional and physical needs of the person, but now the human voice will also be used to put back into the body what the body needs. This revolutionary concept will change the way people use sound as a healing modality for all future generations.

- Dr. Alfred Tomatis found that the voice is a major contributor to the overall stability of the body through sound.

- Dr. Tomatis's Listening Posture is important to the success of a person's listening in general, but is particularly important to the maintenance of the person's improved listening skills.

- Many methods or techniques have developed that use sound, music, the voice, listening, energy patterns, and more to enhance our body's responses in general.

- Human BioAcoustics uses the voice to identify the imbalances, and then through frequency specific sound presentation, re-introduces to the body the specific frequency that will balance the body towards self-healing.

- The voice will only be successful in balancing our body if it provides the correct vibrational stimulation which is accomplished with bone conduction vibration, the correct posture, and knowing which sound is the correct sound to use as provided by the ear.

CHAPTER 6

The Voice-Ear-Brain Connection

*T*he Voice-Ear-Brain Connection is the subtle biological, biochemical, whole body inter-relationship of how the voice reflects and impacts what the ear hears, emits, and processes, and what the brain perceives, interprets, and relays to the body. The entire process is based upon the body's response to sound vibration and is founded in the sciences of physics, biology, and chemistry, and the world of metaphysics. This system provides a checks and balance system for the body. We know if the energy flowing out and coming in is balanced within the system, then the voice, the ear, and the brain demonstrate that balance.

If vibrational medicine attempts to connect the subtle energy fields contributing to the functioning of the physical body, then the Voice-Ear-Brain Connection fits into that category. A cycle of energy exists between our subtle energy level and our physical body: a cycle of sound. It supports the reception and expression of sound for meaning between the voice, the ear, and the brain. To maintain this cycle of sound energy The Davis Model of Sound Intervention was created. This model supports repatterning the energy between the voice, the ear and the brain so that our physical body can be brought to a better natural starting place to learn and develop, and eventually maintain wellness.

How does this happen? Consider what you have read so far. Let's look at the parts.

The System

A system called The Voice-Ear-Brain Connection exists that supports how well the body remains in homeostasis or balance. This system begins at the subconscious level or at the etheric level of the body where the energy patterns are shared within and without the body. Imagine a cycle of vibration (sound) undulating from outside of the body to inside of the body continuously with wave patterns. The entire body feels the effects, including subconsciously. The etheric level sends vibrational support for the functioning of the physical body. And the physical body sends vibrational support through the connective functioning that exists between the voice, the ear, and the brain–the translators and processors of the information. The cycle is not just contained within the physical body. The cycle is only complete when results are only fully realized at the etheric level; the subconscious supporting the conscious awareness of the person. It's a full mind/body experience. The energy patterns can be seen as a cycle moving to bridge the gap between the etheric world and the physical body. The voice, the ear, and the brain are simply the physical forms needed to share the incoming and outgoing sound stability of the body.

By repatterning and stabilizing this energy each person can be brought to a more supportive starting place so that their learning, development, and wellness can be enhanced. This starting place provides a more stable foundation from which to grow. The change typically begins at the subtle energy level and advances to the physical, final outcome, although sometimes a physical injury will start the process in reverse. Within the world of quantum physics the possibility exists that what happens within us can impact what happens outside of us and vice versa. The Cycle of Sound is simply one of responsiveness, the response to the give and take of sound energy. Sometimes the response is not immediately seen, because when change is being made at a quantum quark level the change may need time to be seen at the physical level, or it may never be seen as the body compensates and makes the change before a symptom can manifest.

This etheric or ether level has been associated with sound, the resonating energy holding all things together. Sound is the energy of the universe and everything makes a sound, even inanimate objects. Many scientists and healers have expressed that the universe was created by sound, and humans have only evolved as with the evolution of their human sound, their voice.

All systems ride a wave of ups and downs, ins and outs, or vibrational cycles. Our Voice-Ear-Brain Connection cycle rides the expression and reception of sound at every level of the body, starting at the subatomic level. The vibrations within the external energy field or etheric level affect the vibrations of the physical body, or internal energy field. For wellness to exist, this cycle must stay in appropriate balance.

Within this Voice-Ear-Brain Connection system each person can discover how to remain in balance by using the specific sounds that the body asks for. These necessary sounds are heard within the silence that our bodies crave and need. It is within the silence that we can find our inner self, or our own sound.

The Voice

The voice is key to understanding our overall energy. The voice is our body's own instrument and needs to be in tune. The voice reflects what the ear is trying to express as out of balance. The voice reflects how the body is functioning. If this voice output is out of balance, then the brain is unable to maximize our overall body energy. There will always be something out of tune. The brain needs support in knowing how to make the change or tune up. The brain becomes the conductor and perhaps sends the needed sound to the ear as an emission, saying 'tune your body to this sound'. The brain is responsible for making sure our Signature Symphony of Sound is balanced. The voice will imitate that sound and reconnect that sound to the body, thereby balancing the overall body frequencies, a form of entrainment. The ear receives the new sound by hearing it produced with the voice, and the vocal sound is also the transmitter to the rest of the body by bone conduction vibration. The voice then is our body's stabilizer. This is a continuous cycle that searches for ways to maintain and stabilize our frequency patterns.

The voice alone can be helpful to our healing process, but when sounded with the correct tones, the healing becomes more of a naturalized process, a movement towards wholeness. The voice creates the movement or vibration that moves throughout the physical body, stimulates the mind, and moves outward toward the surrounding environment. This sets the healing process in motion.

The voice will only be successful in balancing our body if it provides the correct vibrational stimulation which is accomplished with bone conduction vibration, the correct posture, and knowing which sound is the correct sound to use as provided by the ear. The voice enhances our overall body functioning.

Within the vibrational levels discussed in this book, sound is important at the etheric level. This level is also associated with hearing. However, the body connector of the chakra at this level is the throat which is associated with the voice. This demonstrates that from other known systems an ear/voice connection has already been identified. The voice is able to express the responses of the other body processes and systems. This connection has been historically noted by many different people, cultures, and religions for hundreds of years.

Subconsciously, we use our voice in some natural way every day: a laugh, a sigh, a groan, and so forth. When in the company of others we are often encouraged to squash using vocalizations because they appear strange to others. However, these natural sounds were and are an important function of an individual's expression of himself and what is happening within him. Human language began as a combination of sounds made by the developing human who attached meaning to those sounds. It is not from the words expressed by the voice that we obtain needed information to stay in balance; it is only the sound of the voice that is needed.

The Ear

Our ear is more than a hearing mechanism. Our ear is our global sensory processor. Yes, we hear with our ear, but all of our senses are vibrationally stimulated through the nerves that pass through the ear. The sensors of the inner ear pick up the subtle changes that provide minute, informational pieces which create an awareness of the world around us. It is only when this subtle sound system triggers a response to the body that our consciousness kicks into action and we communicate either with our own body or with others.

Recently Drs. Sophie Molham and John Foxe, neuroscientists at The City College of New York, were awarded a grant from the National Institute of Mental Health to study whether and how the nervous system integration of different sensory stimuli is impaired for autistic individuals.[1] The premise is that because atypical integration of multisensory inputs has been considered to be a major component of autism, that if they learn when and where in the neural processing stream the sensory integration deficits occur, that they will help define the neuropathology of persons with autism. It is wonderful that a specific project is being undertaken to aid our understanding of autism. With regards to the sensory information received by all people, consider that the ear is the global sensory processor and that by repatterning how sound is received and perceived by the body, this sensory information has the opportunity to relearn how to use incoming information better. The best way that I have found to accomplish this sensory restart is with sound-based therapy, not only for people on the autism spectrum, but for most learning, development, and/or wellness challenges.

Our ear also emits a sound called an otoacoustic emission. This spontaneous outpouring of sound from our ear is providing information about our body's lack of function. Also, this emission is giving us the exact frequency(ies) that our body needs in order to self-heal.

By identifying our spontaneous otoacoustic emission, I was able to connect irregular frequencies with imbalanced frequencies identified by vocal analysis one hundred percent of the time. This became known as The Davis Addendum to the Tomatis Effect.

We need to hear this sound from within. We can only hear this sound in the silence of the body or the environment. We can become internally aware of the vibration of this sound but this takes much training. We need to turn off the external sound world while remaining a part of that world. We need to selectively tune into our sounds from within. The ear then becomes a multi-talented organ of the body.

Many people describe tinnitus as a ringing in the ear. In the past, tinnitus has been defined as "the perception of sound that has no external source".[2] What is the sound? Where does it come from? Many assume the problem is associated with the ear but recent brain research indicates that perhaps the origins are neurological and begin in the brain. A hearing loss can damage the hair cells within the inner ear leading to loss of normal function of the auditory nerve fibers sending input to the auditory neurons in the brain. This loss of normal auditory nerve function triggers physiological changes such as an increase in spontaneous neural activity. This spontaneous activity refers to the firing of neurons at rest when there is no sound stimulation. When there is a loss of normal input the neurons fire at abnormally high rates that is similar to what happens when neurons are stimulated with sound.[3] These same neurons continue to respond to sound but at the edge of the hearing loss. There is a reorganization of the frequency map of the brain. I propose a theory that something similar is happening with our spontaneous otoacoustic emissions. The emission map or graph possibly shows where the frequency map is distorted, even if only slightly.

Recently the theory was proposed that by stimulating the energy of the missing sound on the frequency map of the hearing loss, a form of neural plasticity would reactivate the auditory system thereby reducing the process creating the tinnitus sound. Two studies involving cats demonstrated that tinnitus might improve as a result of acoustic therapies that stimulate the auditory cortex of the missing frequency of the hearing loss.[4] I am proposing that something similar can be done to support the imbalanced spontaneous otoacoustic emission. By reintroducing the sound to the brain, the frequency map is repaired and the brain then supports the body towards wellness. The voice reflects this change.

Most ideas about spontaneous otoacoustic emissions have been associated with one's sense of hearing. The generation and the operation of our spontaneous otoacoustic emissions are not understood. In fact previously, the theory was that a backward traveling wave traveled along the cochlea similarly to a forward traveling sound wave, but at least two papers have disproved this idea.[5] Instead, an experiment at the Oregon Health and Science University showed that the sound waves coming out of the ear travel through the fluid of the inner ear.[6] This fluid appears to be stimulating both sides of the basilar membrane in the cochlea. I am excited by this research because the results demonstrate that the sound is not from sound entering the ear but coming from somewhere else, most likely the brain.

In another study evaluating the response of a medication for tinnitus the author of the study reported, "We're pretty sure the brain is involved. That's important, because for a long time doctors and scientists assumed tinnitus was an ear problem. It was only after they cut the auditory nerve of a number of patients, 'and they woke up and still had the ringing' on top of total deafness, that experts looked to the brain."[7]

The Brain

The brain is our conductor and works diligently to make sure our body's symphony is in correct rhythm, provides the best inflection, maintains the proper duration of sound, and keeps the entire body in tune. The brain turns the codes of the vibrations received into experiences and meaningful responses. With sound frequencies, which for me are all vibrations received in the brain, these codes produce chords, overtones, and harmonics which may only be received by their codes. In many cases, the sound received may not have been heard through the cochlea of the inner ear but heard as a response to quantum energy movements. The brain helps support the body to maintain a level of coherence or balance.

The brain works together with the molecules and biological necessities of the body. The brain has control over the many processes in the body which supports our immune system. The body can mirror a mental or emotional event which may result in an illness. This response triggers vibrational disturbances which will create an imbalance in the Voice-Ear-Brain Connection thereby disrupting any learning, development, or wellness process trying to be established. The energy will be taken away from the learning, development, and/or wellness processes to work on stabilizing or balancing the disturbance. Our body tries to repair irregular patterns.

One example is the many autistic children who appear to have been reactive to their body's influx of heavy metals like mercury, whether it was from vaccinations, environmental, or genetic factors. The child's overall frequency response welcomed the frequencies of the toxin, blending that frequency into their body's functioning. However, the blending of that frequency was not harmonious. The child's body began working to try to protect itself from this toxin and as a result, the child's learning, development, and wellness process were altered. The toxin was not something easy to repel, nor was it easy to assimilate. The substance was like the pea in the story, *The Princess and the Pea*. The pea was the irritant that did not allow the Princess a good night's sleep. The toxin in this case was the irritant that did not allow the body to function well. Once the body began to work incorporating the toxin as a complement to their body, the delays for growth in the child with learning and developmental skills became evident.

I have seen thousands of autistic children during my professional career with sound-based therapy. I began to see a connection to their development and wellness related to histories of vaccinations long before the literature reported such. One case in particular stands out. A child, age four diagnosed with autism came to The Davis Center for an evaluation. I saw many issues present but he was non-verbal and reactive to sound. As an infant, he had started an attempt at language, but his father reported that around fourteen months, the desire to talk ended, and he saw a progression of this happening after a series of vaccinations. I began working with this child at age three. He started with the Roots of The Tree of Sound Enhancement Therapy and made nice progress. His father noticed that he was less sensitive to sound and began noticing that sound around him had meaning. He progressed into the Trunk of the Tree and started a basic Listening Training Program. At the end of the basic program which consisted of sixty hours of listening, his father reported that he was beginning to socially connect with others around him, words were pronounced more clearly, his vocabulary was expanding, his inappropriate behaviors improved, and sounds were no longer sensitive to him.

As is often the case with autistic children, this child had to return for a continuation program to help support him through the changes he obtained as well as help stimulate the enhancement for new possibilities. During the second continuation program the father called frantically one evening. After his son's Listening Program in the afternoon he took his son to the doctor who administered another vaccination. The child immediately reversed the advancements seen up to that day back to almost beginning levels! The father was in tears. The doctor said that the change could not have been from the vaccination, however, the change in the child's reversal was overly dramatic. I had to begin almost from scratch to advance the child. As sound therapy is not covered by insurance, nor accepted within the educational and medical fields due to the position papers I've mentioned, this struggling father could not continue the listening sessions although he knew that our program had been the only program that provided the most significant change for his son.

So this case demonstrates that the vaccination response triggered a vibrational disturbance within the child's body which in turn created an imbalance in the Voice-Ear-Brain Connection, thereby disrupting the learning and developmental process that the child was struggling with. The energy was diverted away from the learning and development process in order to try to stabilize or balance the disturbance, and as a result the child appeared lost in his world of autism. (This does not mean that all vaccinations create this type of change. This is simply one example of how external sources can change internal vibrations and create an imbalance.)

The Body

The body is in a constant give and take of energy, a constant cycle. We express and take in energy. All energy has a frequency, therefore a sound. Vibrational medicine suggests that the body is a dynamic energy system. By manipulating the energy system, the fields are rebalanced, restoring order to cellular physiology. Our bodies are complex networks of energy fields at the physical level: circulatory, blood, and nervous systems to name a few. The Voice-Ear-Brain Connection is a subtle energy network linking the deepest level of the person – their soul – to the physical body. Our physical body is only visible to ourselves and others because our conscious mind defines our vibrating structure as a physical form. Our subconscious mind understands that there are cellular vibrating structures within us that create the conscious shape and directly focus the energy along certain pathways, i.e. between the voice, the ear, and the brain. And our subconscious mind also understands that those vibrations are sending energy to the vibrating universe around them in order to make basic choices for survival and living, i.e. "Do I run from an approaching bear?" The Voice-Ear-Brain Connection exists because of the vibrations of sound energy at the subconscious level affecting the conscious level. In general, our bodies can self-heal because of the communication within our cycles. The cycle is only able to function within its existing energy. If a disturbance is present and the energy is distorted, the response will then be distorted, irregular, imbalanced, or dyssynchronous. No change is possible without a change in the energy pattern.

Because the Voice-Ear-Brain Connection incorporates the physical body, the emotional and mental processes are also stimulated. For some, no change is possible if the emotional or mental blockage isn't removed. Sometimes the vibrational energy taps into those deepest levels and releases them. The voice will then reflect the release of that change and connect with the brain allowing the change to reach deeper consciousness. A new level of inner awareness supports a better flow of energy, resulting in feeling more balanced.

The Voice-Ear-Brain Connection is biological and biochemical because all cells, all processes, all substances used by the body have Frequency Equivalents, meaning they have a particular frequency number that is present on a mathematical matrix of predictable frequency relationships, as shown in the science of Human BioAcoustics. And communication between and among the cells are within organized patterns of vibration. The whole body connection is evidenced by the vibrational sensations sent through our body systems like the circulatory and nervous system. The vibrations are waves of sound energy triggering responses when the correct neurons are activated. All body frequencies are identified through vocal analysis and can now also be said to be identified through spontaneous otoacoustic emissions. Today, it is considered true that a disease can show up in a person's energy field before the symptoms are manifested as a pathology. This can be seen with the connection of frequencies between the ear and the voice.

When Sharry Edwards was in the beginning stages of developing Human BioAcoustics, she reported a story about a salesman who came to her center. He had sold her equipment that she used to monitor body responses during what is known as tone trials, introducing the specifically identified frequencies to the body under supervision. In order to demonstrate to this salesman what her evolving method was, she took his voiceprint and noticed that thyroid issues seemed to be imbalanced. She queried the salesman by asking him if his thyroid had ever caused him problems. He said no. On a doctor's visit soon afterwards, he mentioned this to the doctor who offered to do thyroid testing to double check. The results did not indicate any difficulties. So the salesman, thinking Ms. Edwards' process was nothing special, went on a sales trip. He was in a very rural area of the country when he collapsed and was rushed to the hospital. Many tests were administered and just before he passed into a coma, he asked them to check his thyroid. The staff did and told him afterwards that the thyroid testing probably saved him. It was the exact problem that needed to be identified and from that diagnosis they were able to help him get better. Ms. Edwards had been able to identify an evolving problem through vocal analysis fourteen days in advance of the problem manifesting symptoms! Other cases like this have been reported but this one always stood out dramatically.

The Foundation behind the Voice-Ear-Brain Connection

This system would never have been identified without a series of five laws known by its parts: 1) The Tomatis Effect and 2) The Davis Addendum to the Tomatis Effect. Part one has been in existence since 1957 when Dr. Alfred Tomatis demonstrated to the French Academy of Sciences a phenomenon that he noticed after he worked with many different populations. He observed that the voice only contains the harmonics that the ear can hear and if you provide the correcting sound to the ear, the ear is restored to a full complement of sound. If the sound is maintained over time, the voice regains its natural, balanced abilities.

The Tomatis Method has promoted this effect for the last fifty years and there are many spin-off methods that have been developed. Most new sound-based programs say they are Tomatis-based but the program can't be Tomatis-based unless they include the full process of the Tomatis Method. It's like saying you know algebra because you can add two numbers. The tenets of The Tomatis Method are much more complex than many of the spin-off or Tomatis-based approaches.

The Tomatis Effect and subsequent Tomatis Method show the connections between the ear and the voice. Those two parts being in balance are important for learning and developmental skills to evolve or remain intact. Thousands of people have been helped with this method around the world.

Dr. Tomatis showed that the listening ability of the ear was connected to the vocal patterns. By retraining the ear to better listen to the frequencies received, the voice has better quality, rhythm, pitch, melody, and intonation. The voice is richer.

The Davis Addendum to the Tomatis Effect or part two identifies a further connection between the ear, voice, and brain. By demonstrating that the ear and voice contain similar imbalanced frequencies when those frequencies are broken down into brainwave multiples, it reveals how the brain works by the mathematical connections within the harmonics of sound.

The Davis Addendum demonstrated that the output of the ear, called spontaneous otoacoustic emissions, can be measured in specific frequencies and that by matching these emissions to the specific frequencies of stress in the voice that an additional connection between the ear and voice is present.

All five laws incorporate the concept that by improving what the ear processes, either as an intake of sound or expression of sound, the voice becomes coherent or balanced. This coherent voice then demonstrates fewer irregular patterns. The result demonstrates the attempt to heal the communication between The Cycle of Sound and the sense of self. The feeling of being one with who we are and how we fit into the world comes together.

This process is a whole person approach because the balance of the energy signals within the body is more important than any physical symptoms, which typically show up only after the signals are imbalanced. Balancing the vibration of our sound signals is central to our improvement, and is central to being able to maintain our ability to learn and develop, and maintain a state of health and wellness.

Evolution of a New Process

The Voice-Ear-Brain Connection is a previously unrecognized system of the body. Its importance is major to each one of us. Understood with the five laws of the Tomatis Effect, The Davis Addendum to the Tomatis Effect, and demonstrated with the study of quantum physics, the subtleties of this unrecognized system can be brought into our consciousness and allow our bodies to evolve to a place of equilibrium. Perhaps the internal body disruptions and external influences can be minimized when this Voice-Ear-Brain Connection system is understood better. The body will work to adjust the energies so that the system works harmoniously with the body and its other systems. This cyclical connection is major to the development of self. The process supports our understanding of our personal identity and our connection to those around us.

This newly discovered and important system within each one of us begins in utero. Those connections initially started in utero need time to integrate themselves into the person's body functioning. Any daily impact, whether it is physical, emotional, or environmental can directly challenge the stability of this connection. There is a mutual dependency between its parts: the voice, the ear, and the brain. Without the parts the whole will not function as a balanced system. The dependency is based upon the cyclical relationship of the expression and reception of cellular signals as sound within and without each one of us. When in balance the parts work powerfully together. When unbalanced many opportunities for disease, disorders, or discordance exist. The cycle reflects a system of responsiveness between the energy levels within and without the body.

Many clients leave my office asking my front staff if I have psychic abilities. I can often tell how early childhood impacts the person based upon one of the tests that I administer. I may say, "What happened at age two that impacted your son?" Answers may run the gamut from, "That's when his sister was born", or "That's when he began a series of middle ear infections that just wouldn't go away", or "That's when my mother passed away and it took me awhile to pull it back together". These everyday life experiences impact people in different ways. We often are unaware of the impact at the time, but because emotionally we can hold onto an event at the cellular level they have a way of impacting us for a long time. The issues can be seen during childhood with something as severe as selective mutism, or less severely but still impactful, as lack of attention. For teens, the issues can rear their head as social disconnections, reading challenges, or drug dependency. For adults, whether young or older, the issues can be seen in mood swings, depression, dyslexia, severe illness, inability to maintain a job, difficulty with marital relations, and so much more. It is important to release these emotional and/or physical responses, and help the person move forward towards a more balanced person. Sound-based therapy can accomplish this.

The Voice-Ear-Brain Connection integrates the whole person, not just the parts. The balance of the entire system is more important than the balance of the parts. The same applies to the use of sound-based therapy. The entire interpretation of The Tree of Sound Enhancement Therapy is more important than any one part of the Tree. The use of one therapy may not provide the needed change, whereas a combination of therapies as defined by the Diagnostic Evaluation for Therapy Protocol may produce the desired changes since the entire Voice-Ear-Brain Connection has been addressed.

The Davis Model of Sound Intervention was designed to address the whole person. The concepts behind this approach stem from three factors:

1) The five laws of The Tomatis Effect and The Davis Addendum to the Tomatis Effect;
2) Every cell in our body emits and takes in sound as frequencies; and

3) The ear is more than a hearing mechanism. It is your global sensory processor because of how sound energy travels through the central nervoussystem, connective tissue, and skeletal system as stimulated through the ear and the body.

These three concepts bring clarity as to how sound impacts the entire body. When the energy is unbalanced in either number two or three above, then number one is disrupted. Number one provides the foundation for acknowledging a connection between the voice, the ear, and the brain. Number one reflects The Cycle of Sound. The other two concepts provide the balance needed to stabilize the process.

Prior to this book, the only way to balance this Voice-Ear-Brain Connection system was with the use of sound-based therapies as defined by the Diagnostic Evaluation. But this process wasn't readily accessible due to its limited availability, plus there was the desire for developers of certain therapies to only push their particular therapy. The first limitation is now lessened because the test battery can be administered in out-reach locations and because the process is currently under consideration to be streamlined. As to the second limitation, business development will always be a factor when considering any therapy.

Sound-based therapies are still needed when a system is out of balance and all persons are encouraged to start with the Diagnostic Evaluation for Therapy Protocol to determine if any sound-based therapies are needed to bring the body into balance. The changes within the body and the techniques learned during the therapies support the necessary skills to benefit in the best possible way from Ototoning.

Ototoning is the missing piece needed to totally restore a person to maximum homeostasis. No more "Let's get together and tone", or "Try this mantra". Even after consideration of sound-based therapies used in the appropriate order, Ototoning will provide the tool for maintaining and supporting the evolved changes within the person from the therapies. Ototoning provides the opportunity for each person to know what their body is telling them it needs, them alone, not their neighbor or someone across the room. Ototoning bridges the gap between sound-based therapy and other sound interventions, and provides a way for the body to naturally entrain to our needed sound. Ototoning is a natural self-entrainment technique.

Sound Notes
Chapter Six Summary Statements

- The Voice-Ear-Brain Connection is the subtle biological, biochemical, whole body inter-relationship of how the voice reflects and impacts what the ear hears, emits and processes, and what the brain perceives, interprets, and relays to the body.

- The Davis Model of Sound Intervention supports repatterning the energy between the voice, the ear, and the brain so that our physical body can be brought to a better natural starting place to learn and develop, and eventually maintain wellness.

- A newly discovered system called The Voice-Ear-Brain Connection exists that supports how well the body remains in homeostasis or balance.

- The voice is key to understanding our overall energy. The voice reflects what the ear is trying to express as out of balance.

- The voice creates the movement or vibration that moves throughout the physical body, stimulates the mind, and moves outward toward the surrounding environment setting the healing process in motion.

- The voice will only be successful in balancing our body if it provides the correct vibrational stimulation which is accomplished with bone conduction vibration, the correct posture, and knowing which sound is the correct sound to use as provided by the ear.

- Natural vocal sounds are important for mankind's expression of himself and what is happening within him.

- Our ear is our global sensory processor, because all of our senses are vibrationally stimulated through the nerves that pass through the ear.

- The spontaneous outpouring of sound from our ear is providing information about our body's lack of function, giving us the exact frequency(ies) that our body needs in order to self-heal.

- The sound received by the brain may not have been heard through the cochlea of the inner ear, but heard as a response to quantum energy movements.

- By manipulating the sound energy system the body fields are rebalanced, restoring order to cellular physiology.

- The Voice-Ear-Brain Connection would never have been identified without a series of five laws known by its parts: 1) The Tomatis Effect and 2) The Davis Addendum to the Tomatis Effect.

- All five laws incorporate the concept that by improving what the ear processes, either as an intake of sound or expression of sound, the voice becomes coherent or balanced.

- Sound-based therapies are still needed when a system is out of balance, and all persons are encouraged to start with the Diagnostic Evaluation for Therapy Protocol (DETP) to determine if any sound-based therapies are needed to bring the body into balance.

- Ototoning is a personalized toning session based upon the sound that the body tells them they need, and bridges the gap between sound-based therapy and other sound interventions.

CHAPTER 7
Ototoning

*D*on Campbell in his book, *The Roar of Silence: Healing Powers of Breath, Tone & Music,* shares advice from African musicians:

> "Sing the song of the soul. Let it rise from tone. Let it rhythmically flow into its own life. Let it become You. To sing another's song is to rob the other's power and your own. Move your own body to its rhythm. There is danger in singing someone else's song. The unnatural spirits will enter your body then. Just sing yourself over and over... you'll be purified...
>
> Tap the powers that lie beneath the consciousness of the song. The song leads down to the chant, the chant leads down to the tone, the tone leads down to the breath, and the breath leads to the energy beneath the sound – the roar of silence." [1]

It is within the silence that we can become connected with our own song. We can introduce our own song back to ourselves with Ototoning.

Some scholars suggest that Sanskrit is the oldest language on our planet. There are approximately ninety languages in our world and many have words derived from Sanskrit. The English language has about twenty-five per cent of its vocabulary derived from a Sanskrit base. While Sanskrit has an abundant literature, poetry, philosophy, and drama, some scholars feel that it is a foundation for a global language of consciousness, modern science, and spiritual science. While most languages have rhythm, harmony, and poetic sounds, Sanskrit has these pieces and more. In fact to some, Sanskrit is the only language available that has the letters and sounds that can be used to produce powerful mantras. Sanskrit alone has the power of the divine sound needed for a mantra.[2] However, even with this powerful impact, the final piece for each individual to feel his own sound will only come from a tonal sound that each person's body wants and needs to feel whole.

Silence is necessary for sound to be perceived. Within the Sanskrit tradition the Anahata Nada means a sound not made in the way we know it. This sound is the sound of the universe, or the primal sound of all energy. According to the Vedas, Aum or Om is the most sacred of all words because from Aum the universe evolved. By itself, Aum is not the Anahata Nada. The sound is composed of four parts: The individual vowel sounds of the 'a', 'u', and 'm' and the silence that precedes it and appears at the end of the sound combination. The silence supports the three individual sounds. Within this tradition the production of the 'a', 'u', and 'm' is not the objective alone. The production of the silence or unmade sound supports the toning sound.[3]

There has been a connection understood between sound and well-being for centuries. To be in sound health means keeping the body in a state of wellness. Tonics were made as a way of keeping the body invigorated or to build up strength. The word *tonic* in music refers to one's tonic note, the central pitch around which melody and harmony are combined and balanced. The tonic note has also been referred to as a fundamental note. Within Ototoning this fundamental note refers to the note that supports one's own natural self-healing. Ototoning is the ear/voice tonic for keeping us healthy.

What is the difference between music as medicine and sound supporting wellness? In a recent MSNBC release about music as medicine the reporter mentioned that a rising number of hospitals in the US are including music in the treatment of many ailments. Doctors themselves are studying and using music to change the body's neurons and blood-carrying cells. The way this press release discussed the impact of music to the body is:

"The anatomical route musical notes take through the body is indeed a busy highway celebrated in many songs, from head to heart. Based on interviews with neurologists and cardiologists, the journey from an instrument string to your heartstrings goes something like this:

Sound waves travel through the air into the ears and buzz the eardrums and bones in the middle ears. To decode the vibration, your brain transforms that mechanical energy into electrical energy, sending the signal to its cerebral cortex – a hub for thought, perception and memory. Within that control tower, the auditory cortex forwards the message on to brain centers that direct emotion, arousal, anxiety, pleasure and creativity. And there's another stop upstairs: that electrical cue hits the hypothalamus which controls heart rate and respiration, plus your stomach and skin nerves, explaining why a melody may give you butterflies or goose bumps. Of course, all this communication happens far faster than a single drum beat."[4]

The release also mentioned that classical music was the typical pick of physicians and therapists who use melody as a healing tool. This statement alone helps define the difference between music helping to heal and sound helping to support the body with its own wellness. The music approach uses the melody of a musical selection. This melody helps entrain the person's body rhythms, and promotes physiological change in the body cues to support healing. The process used is through the auditory function of the ear. A sound approach can't help but include the auditory function, but the sensations address more than audition and the rhythms and melodies of music. Sound goes to the core frequency energy of the body meaning in more depth than music per se. Even though music is based upon mathematics, the sound mathematical specific frequency responses within the body are often more than a melodic impulse or response.

Stephanie Heidemann, a singer, songwriter, and authentic voice coach said it beautifully, "Sound is the frame for silence, which is the healer. Music is the dance between the two."[5] Music only becomes one part of The Cycle of Sound. As Ms. Heidemann uses toning and mantras within her approach, she stresses that after a period of internal toning the silence after the tone helps the body adjust to the vibrations received and produced.

To be sound in body, mind and spirit, the body must be in harmony. The frequencies of the body must be in tune. To be in tune one must be able to tone. Toning and mantras have been used for thousands of years, but until now the sounds used were universal sounds – sounds that all people can respond to. While harmonious sounds can balance the body, so too can discordant sounds trigger disharmony. While toning, chanting, and mantras work on balancing the body responses, the more specific technique of Ototoning addresses the specific needs of the individual, not the individual within the group.

What is the difference between these types of soundings? Toning uses the repetition of a single vowel or syllable. Typically the vowel is preceded or followed by a consonant. The sound is held for a period of time and may follow the rhythm of the person's breathing. When the vowel sound is held the sound is determined by the shape of the resonating spaces within the throat and mouth. Additionally, the sound is enhanced by the overtones produced in the sinus cavities. When done strongly, toning can alter vibrations at our core cellular structure. Toning is thought to be a way to restore people to their natural harmonic patterns. Toning in most instances is considered a right-brain activity because it uses a creative, expressive, and spontaneous sensation of sound and feeling. However, when specific vocal intent is introduced the activity becomes more whole brain.

Mantras are sounds implanted in the mind, often as a catalyst for ridding the self of an impediment and follow a rhythm of repetitive sounds. The sounds, which are often syllables, words, phrases, or longer hymns, can be said out loud or inwardly within the mind. Mantras require concentration by the chanter. Chanting is a type of singing following the person's breathing patterns using short tonal phrases within a narrow frequency range. They often are associated with a spiritual setting. For example, within the Muslim religion chants are a part of daily life because chants are the individual's call to prayer, or are used as a celebration for religious life. Both of these sounding techniques can be vocalized with others around, and each can be shared with another person. A group setting is okay for either of these techniques. These two activities incorporate more left-brain activity than toning typically and are whole brain activities.

To be in total harmony within your own body it is important to find the sound that your body needs, not what your neighbor needs. Toning should include finding out what your own body wants. Find it within yourself. Attune to your own body, mind and soul – your own physical, mental and spiritual needs. The way to this harmonious level is with your own needed tone. This can be accomplished with Ototoning.

The other sounding techniques like toning and mantras connect the person to the universe through a universal sound, Aum. When produced, one is connecting the person to the sound of the universe. It is often the overtones or harmonics produced that activate the vibrational energy of the tone. These overtones or harmonics are inherent in every sound we make and create the individuality of the person's voice or sound. Mantras, although they can be individualized, are often done in group sessions. The origin of mantra was to have the person create a soundless sound while deep in silence. The intention put into the mantra is important. Chanting done in a group setting such as a sanctuary is powerfully received because of the effects of the acoustical qualities of the sanctuary on the person. These techniques have been used to connect the person to the universe with their fundamental tone. Often the tones used are whole notes or combinations of whole notes. Notes such as the F, A#, B, or C are generic so to speak.

With Ototoning it is more important to find your individual sound and vocalize it back to the body. It is not necessary to warm up the voice using vowel sounds, incorporating the chakras, using loudness, pitch, and timbre; you just use your own sound. Ototoning uses the spontaneous sound of a hum produced vibrationally through the entire body with bone conduction vibration matched to the sound or tone emitted from the ear. The notes vocalized may or may not be generic notes. The sounds are specific to what the ear emits. The sounds coming from the ear as demonstrated by The Davis Addendum to the Tomatis Effect are directly connected to the voice.[6] Vocal analysis has demonstrated that we each have a Signature Sound.[7] The correct and needed sound is vocalized back into the body by Ototoning.

Consider the results of some early studies involving spontaneous otoacoustic emissions. Tinnitus is often described as a ringing, buzzing, or sound heard within the ear by the person. In patients with spontaneous otoacoustic emissions induced tinnitus, by suppressing the spontaneous otoacoustic emissions the annoying tinnitus sound was eliminated. Over the last ten years devices have been developed and put on the market successfully that identify the frequency of the sound the person hears and re-introduces that sound back to the person's ear, and over a period of time, typically eighteen months, the person's tinnitus sound disappears. This product found a way to stabilize and eliminate an annoying sound by reintroducing the sound to the body with a device. With Ototoning the sound the person wants or needs is reintroduced to the body with the voice, which vibrationally is a faster approach.

Ototoning addresses the person's needs for their inner self and works towards stabilizing that sense of self or their inner soul. Once the body is stabilized then one can advance to sharing their energy with the universe. The individual must be internally in balance before connecting to the external environment. Identifying the specific needs of the body supports the person towards a maximally balanced body instead of generically balancing a person based on sound techniques not specific to the body's needs. Consider the impact on the universe when more people are better connected and balanced individually; the universe's energy will be even more significant!

Meditation is thought to connect us to knowledge learned from our higher consciousness. It builds a subtle energy bridge between the two levels or bodies. There are varying degrees of meditation, but the one mentioned here deals with sound. For thousands of years repetitive sounds have been used to vary the rate of consciousness. For some, a mantra repetition has helped to clear the mind of conscious thought. Mantras help the person reach new spiritual levels and chanting helps people stabilize their body rhythms.

Meditation has been shown to activate a system of resonantly tuned oscillators powered by the circulatory pulses of the heart. The heart can also be powered by one's breath stream. The heart rhythm can be changed during meditation, which in turn creates a resonant vibrational link between the heart and the brain, especially the sensory cortex. Meditation has been used to release stressors stored in both the physical and ethereal body by many people.

In sharing information on the primordial sound meditation technique, Deepak Chopra writes:

"From the perspective of the sutras, or Vedic sounds, there must be a distortion in the proper sequence of intelligence (of the cellular force fields – author's addition) as it unfolds into the relative world. ' Hearing' the virus in its vicinity, the DNA mistakes it for a friendly or compatible sound, like the ancient Greek sailors who heard the siren's song and were lured to their destruction. This is a believable explanation once one realizes that DNA, which the virus is exploiting, is itself a bundle of vibrations.

If this explanation is valid, then the remedy is to reshape the improper sequence of sounds… These sounds are basically like pottery molds – by placing the mold back over the distorted sequence, one guides the disrupted DNA back into line. This treatment is subtle and gentle in its effects… Once the sequence of sound is restored, the tremendous structural rigidity of the DNA should again protect it from future disruptions."[8]

Consider the sounds or vibrations of the DNA as perhaps the otoacoustic emissions that one hears in our silence. This sound is used within the primordial sound meditation. Is it any wonder then, that Ototoning which includes a vocal sound determined from one's spontaneous otoacoustic emission, and is slightly different from a mantra or chant, can produce a meditative state? Or that the re-entered sound balances the overall harmony of the body? The subtle energy bridge connecting the etheric level within the Voice-Ear-Brain Connection with the physical body is stimulating, enhancing, and enlightening the person's responses. Ototoning reshapes the improper sequence of sounds within the body.

Many people feel that the reason to meditate is to achieve enlightenment.[9] When deeply into a session, Ototoning can also achieve a high state of awareness. Ototoning helps the individual produce a higher level of perception and connects the person to a higher consciousness because the body tells itself what it needs. The ear produces the sound it needs for the voice to vocalize and energize the entire body. Once energized, reaching this higher level of awareness becomes easier and more impactful. The ear provides the sound necessary to balance the body's overall combinations of sound. Ototoning can tune up one's Signature Symphony of Sound.

Barbara Bliss in her doctoral thesis entitled *The Singer and the Voice*, where she compares the development of a singer's voice and the development of their psyche, says,

> "Elementary breathing and humming exercises done in a meditative fashion can permit the voice to emerge in acceptable coordination with the personality and the psycho-somatic climate of the moment."[10]

She goes on to say that learning to sing with one's authentic voice is like developing the wholeness of the personality. This example of using breathing, humming, and voicing to balance the psyche can be compared to the sensation aroused with Ototoning. One feels whole, sound, and alive because the person's own breathing combined with humming the necessary sound brings one into balance. Breathing may provide the breath of life, but humming supports the sound vibrational energy to resonate within and without the body. A person's vibrational energy remains after the breath of life stops.

Establishing the best connection between the voice, the ear, and the brain is very important for Ototoning. If we consider that each cell in the body is an instrument emitting its own note, then it is important to keep the instruments in tune daily. Novalis, the poet and philosopher (1772-1801) wrote: "Each illness has a musical solution. The shorter and more complete the solution, the greater the music talent of the physician."[11] Why not use the talent of the person to self-heal? The music needed to heal comes from within the body. The sound is short and repeated over time. It is a complete solution to our own body's needs. Our own health comes from being in balance and in harmony, or in other words, being sound.

I recently had a scare with the possibility of lung cancer. A nodule in my left lung was found over two years ago during an x-ray for a broken rib. I had it monitored after six months and it hadn't grown which was good. On a recent routine physical, my physician suggested that I have it retested and this time it had tripled in size. I was told to immediately go in for surgery to remove the lower third of my left lung. I saw one thoracic surgeon who said to wait for six more months since the nodule was small. Then I went to one of the top thoracic chest surgeons in Manhattan who also wanted me to have surgery, but he wanted tests first. Because of the size he suggested bypassing certain tests and going directly for a biopsy. To me that is an invasive test and I wanted to try other possibilities first. We agreed that I would have a PET scan to determine if the nodule was cancerous. However, I was warned that this may be a false negative if nothing came up because of the size of the nodule.

As soon as I heard that I had this nodule, of course, I began my own sound work, but I also did numerous other treatments: acupuncture, lymphatic drainage, Bioresonance, and Biofield resonance, among others. Plus, I changed my diet to an almost vegetable diet but I also eliminated wheat, gluten, and dairy. I went to a resort health spa and made sure I was taking care of myself. After all, everyone kept saying that the healer has to take care of herself! I heavily worked with the sound-based therapies available to me, especially BioAcoustics. And then one day my son said to me, "Mom, have you been doing your own work of Ototoning?" Well, not as much as I should have been. So I began intense sessions of Ototoning and began to feel repatterned.

After one of my sessions, my friend Don Campbell who is the author of *The Mozart Effect*, called to chat and I related what I was doing. I knew Don had helped himself reverse some wellness challenges that he had with sound, so he was a good friend to discuss what I was doing. He suggested also visualizing in color while I was producing the Ototone that my body was requesting for the day.

Over the next few days I began the Ototone process and incorporated color into my visual view (my eyes were closed). During the first session I saw red heavily with slight orange at the end of the session. In the second session I began with red and moved quickly into orange, but then yellow became very bright. For the third session I started with red, and moved quickly through orange and yellow, and then saw green. It was a calming green. The next session proceeded with green and moved into blue.

Soon after this, I was in a meeting with my staff and reported on my exciting Ototoning sessions. The sound was making a powerful shift for me! I related the colors I was seeing, and one of my staff said, "Dorinne, do you realize you just evolved through the colors associated with the chakras?" I hadn't realized it at all, but I should have thought about it. The blue had brought me to the self-expression level of the fifth chakra, the throat chakra. I began to share more of what was happening to me with others. I was more freely able to express my concerns by using my voice. My Ototoning sessions were very powerful!

Color therapy is not a new concept in the twenty-first century. Today, esoteric color theorists connect the colors of light in the visual spectrum to the lower octaves of higher vibrational energies, which then contribute to one's auric field and subtle energy body. Each major visible color has qualities linked to a chakra. Although outside the scope of this book, the colors associated with what I was seeing while Ototoning can relate to my body's energies and health. And once the highest level of the color's of the chakras are reached with the Ototoning, then I will be able to move even more deeply into a meditative state that should also enhance my overall response to the process of healing.

In the meantime, I put off the PET scan for three weeks while I was hoping that all the alternative practices that I was doing would help in some way. I knew after my Ototoning sessions that I had made a change but had no way to measure it. The PET scan revealed that the nodule was slightly smaller than it had been three months previously, and the nodule had not shown up as a hot spot or cancerous. Since the results may be false negative I will repeat a CAT scan. I am continuing my dietary changes, sound frequency work with BioAcoustics, and Ototoning, and will continue with some of the other alternative methods as well.

The results were accomplished based upon the connection between the voice, the ear, and the brain. The ear provided the sound that I needed, my voice produced the sound I needed back into the body, and the brain made the necessary changes for a more harmonious body. The colors visualized showed the evolution of the healing process through another medium.

As you are learning, the voice, if in balance can be cathartic and can be used to cleanse ourselves from any internal disturbances. Why do we enjoy making sound or singing? Because we can create this catharsis. When we sing or hum, the sound vibrations resonate into our sinus cavities as well as our bones. The sound fills our head with the motion of the sound wave and the sound itself. Our brain receives the sound and processes it as the mathematics of that sound: the chords, the harmonies, the tonal qualities, the melodies, the inflections, and more.

Here is another instance of how Ototoning has helped me. Yesterday, I saw the dentist. I am removing some of my old mercury fillings since amalgam mercury has come up in a lot of my energy readings as imbalanced. I mentioned to my dentist about the nodule in my lung and how, due to my dietary changes, sound work and other energy work, I felt I was doing remarkably well. During the dental work I felt as though I was reacting to two things: aluminum chloride and methyl methacrylate. There was a definite response. Breathing was difficult, my throat was constricting, I was producing lots of phlegm, and I started coughing like the old days. I was having a reaction that triggered my lung and nodule to act up which was concerning to me as I was having my repeat CAT scan soon. And to top it off I developed a whopper of a headache! After returning home I used some sounds for BioAcoustics, but was too ill to do an analysis. Then it dawned on me, why not Ototone? As I was doing it I couldn't see any color, everything was black! The black was most likely the toxins in my body from yesterday. Eventually, the black became strips like band aids embedded on a green background. My voice was really weak and sounded very strident. I felt the Listening Posture immediately support me as more cavities opened up within me for the sound to resonate. The sound became even more powerful; I was feeling a positive response all over my body. Then I noticed my headache went away. Wow, I love this technique!

Don Campbell in his book *Creating Inner Harmony: Using Your Voice and Music to Heal,* emphasizes using "the voice as a gateway to the alignment of mind, body, and spirit".[12] Although he uses the voice in his book to release tension, soothe the mind, and support better breathing patterns, the voice is taking the body's out of tune symphony and trying to make the instruments play better. What the body needs is the correct frequency so we can tune ourselves. This is the keynote or the note emitted by the ear. In this way we get our Signature Symphony of Sound to play beautifully.

Toning is the elongation of vowels. One takes a vowel and holds onto the sound, breathing through the production of the sound, getting the mind to feel the sound, and becoming exhilarated with the sound. The sound is typically not a loud sound, just one that the voice enjoys making and the body enjoys hearing and feeling. The vibration becomes like a massage of sound all over the body. Toning has been used for thousands of years to help integrate one's sense of self. Toning has been described as integrating the outer world with our inner tone.[13] But Ototoning balances the inner world where the true self resides so that the outer world's impact on the body is not as disordered or chaotic.

Toning by itself gives rise to a powerful energy flowing throughout the body. The smallest vibration of tonal sound can stimulate the body's cell structure to release tension, move molecules, and/or stimulate emotions. Body disharmonies can be repatterned with tonal vibration. Toning typically uses generalized vowel sounds but Ototoning utilizes the specific sound the body is saying it needs. By knowing which sound the body wants and needs, one's voice provides the flow of sound – the give and take – back into the body, in a cyclical pattern. The sound moving out of the body is then balanced back into the body by one's voice.

Toning is a conscious activity. The toner feels alert and in control of themselves and the world around them. One can experience a new awareness of life and a new connection with the universe. Imagine how much more powerful this experience can be if the toner provides the sound that makes the body whole or balanced with what it says it needs. One then becomes an Ototoner, not just a toner. One can connect and balance themselves directly to their body's needs and feel balanced, so that when they advance to connecting with the universe they are already fulfilled and balanced at their core or soul level.

Ototoning incorporates the vibration of sound through our skeletal system. Bone conduction vibration is a powerful way to feel the sound throughout the body. Picture the bones of the body as the sounding board of a piano. The sounds are resonating throughout the cavities of the body with bone conduction vibration. The Reverend Ted Karpf writes in the foreword to Don Campbell's book, *Sound Spirit: Pathway to Faith,* "that the music of the spheres is in our bones before it reaches our minds."[14] How true that is!

Dr. Alfred Tomatis studied the connection between the ear and the voice. He thought the ear was like a human antenna. The antenna needed to be tuned daily through movement, listening, and stimulating the voice. He introduced bone conduction humming as an effective way to stimulate the body. He also used Gregorian chant within his Tomatis Method. He felt that the vibration created by this humming or chanting could tune the body to a spiritual resonance. With his technique he felt that the soul would recover its essential vibration and the body its perfect rhythm, thereby bringing the person back to his original state of being. He also felt that the bones and skin of the body were like a third ear. Vibration felt with the third ear spanned from the top of the skull to the bottom of the feet. Bone conduction in particular can help the body feel alive. Ototoning incorporates bone conduction because of the speed by which sound travels through the body and because of the powerful impact felt by bone conduction vibration.

Another variable that supports Ototoning is our ability to produce overtones. Each note or frequency, unless it is a pure tone, has a fundamental frequency or specific vibrational rate. Overtones are sounds higher than the fundamental frequency which are produced in a series of pitches above the note – vibrating differently than the note – yet in a mathematical relationship with the initial note. Overtones help distinguish one voice from another. They determine the quality and character of the individual sounds produced and heard. They shape the texture, timbre, and color of the sound. Overtones help shape the individuality of the person's voice. If someone doesn't know how to create overtones with their voice, Ototoning can still be done. However, producing overtones provides a richer experience.

Overtones and harmonics are often thought of as synonymous, but although each harmonic is considered to be an overtone, each overtone doesn't have to create musical harmony. The body doesn't care about the musical harmony when it is trying to self-heal, such as with the sound provided to me when the toxins were present during an Ototoning session. The Greek root *harmos* in the word harmony means fit together, and harmony within our voice supports a blending together to make the body feel good. Listening and producing harmonics of sounds can entrain the brain so that we are stronger in our responses. We have strengthened our weaker frequency responses. But sometimes, the overtones may produce added change because of the mathematical connections that the body needs. Sometimes, it's the discordance that the body likes and needs to balance itself.

Additionally, when Ototoning, if we begin to resonate with the sound the body needs especially with bone conduction humming, the sound disappears. This in effect is a form of entrainment.

Jonathan Goldman in his book, *The 7 Secrets of Sound Healing,* suggests a formula of Frequency + Intent = Healing for a way to have dramatic effects with self-healing using sound. But he additionally adds another formula of Vocalization + Visualization = Manifestation.[15] With his second formula, a person uses his voice and imagines something or someone creating an outcome. The story I shared about Ototoning and the use of visualizing colors definitely shows the power behind his formula. By using his formula with Ototoning some people may have an even greater impact on the session as I had.

The ultimate of any mind/body work is to have the body begin to recognize when to self-regulate. So too with The Voice-Ear-Brain Connection. The sound-based therapies heavily integrate the connections between the voice, the ear, and the brain. Once the connections are supported and working well, Ototoning can then produce powerful continuing supportive changes for each person. But, as with the highest levels of yoga where the yogi hears and produces the silent sound in his brain and body, so too with Ototoning. The highest level is when the person first listens for their needed sound and then produces the vibrations of the particular frequency silently in their mind, as well as through bone conduction. The mind is creating the necessary vibration to self-harmonize the body. The body is recreating the necessary vibration for the body to make change.

This highest level can only happen once the person has experienced the vibration and sensation for a period of time. The memory of the experience must be set in the brain. By performing Ototoning every day for short periods of time, the process will be set in place to do this naturally. The key to success is being able to listen in your silence for what your body is telling you, and then introduce the sound vocally through vibration throughout your body. The silence that precedes Ototoning is the most important part. One must listen for their own sound in the quiet of their own space in the world. For it is their own sound that provides what the body needs to begin a healing process. Their own sound provides the key to making self-change. It is only within silence that one can hear this sound. Once the sound is introduced then the brain knows what to do.

The Process of Ototoning

Ototoning incorporates the connections between the voice, the ear, and the brain. Ototoning is a personalized toning session based upon the sound that the body tells them that they need. In order for the process to work effectively the person should be in balance as demonstrated by the Diagnostic Evaluation for Therapy Protocol (DETP). This evaluation provides information about the body's functioning for the maximum reception and expression of sound, and determines if, when, how long, and in what order any or all of the many different sound-based therapies should be appropriately applied.* It is possible that you are in tune, but if you are not, then doing a sound-based therapy will enhance your reception and perception of sound energy. This enhancement will make the Ototoning session more powerful.

Before starting Ototoning each person should:

- Balance the body with sound-based therapy for maximum reception and expression of sound. This is ensuring that the Voice-Ear-Brain Connection is functioning well and supporting maximum balance;
- Learn how to use bone conduction sound energy stimulation;
- Learn how to imitate a tonal pitch. This piece is an important component of the Trunk of The Tree of Sound Enhancement Therapy and will develop as the person progresses through a Listening Training Program towards the use of the voice within the program. The tone deaf person should not be tone deaf by the end of their listening sessions;
- Bring the body into alignment and balance. This can be accomplished with the use of various alternative approaches, ones that include energy or physical manipulation. Sound-based therapies can be used as well as we often see changes with the body when a Listening Training Program is implemented or when a spine rollover is used. A spine rollover is introduced with a tone box of the specific frequencies of the muscles supporting the spinal column. The science of Human BioAcoustics has created a tone box specifically for this purpose; and
- Find a quiet spot and listen for your personal silence. It is not helpful to group Ototone because you need to produce your own tone, not your neighbors. You cannot Ototone if someone else is attempting the same thing close by or making sounds close by,

---

* At the time of publication this evaluation is only available by Ms. Davis, but hopefully will be available for others in the very near future. Please contact Ms. Davis at info@thedaviscenter.com to learn how to receive the assessment or visit her website www.thedaviscenter.com.

because you may entrain your sound to theirs. This does not help your needs, but balances theirs.

Following these five steps will produce the best response for Ototoning. Can someone Ototone without one of these steps? Yes, but the effect will not be as demonstrable.

Tinnitus, hearing loss, hypersensitivity to sound, otoacoustic emission testing, and dizziness have all been impacted to some degree by diet and nutrition. So until further research is available, prior to listening to your own sound, consider these precautionary steps to ensure the best possible opportunity to hear or sense your sound:[16]

1) Keep your diet healthy. B[12] Deficiency has been linked to tinnitus in at least one study.[17] Low levels of magnesium have been linked to noise induced hearing loss;[18]

2) Avoid coffee, tea, and other caffeine substances which can deplete magnesium levels, especially within an hour of beginning an Ototoning session;

3) Avoid alcohol which can deplete magnesium levels, especially within an hour of beginning an Ototoning session;

4) Avoid loud noise exposure for at least four hours prior to Ototoning; Avoid aspirin which has been known to create tinnitus, especially for two hours prior to Ototoning. Also avoid foods with salicylates which are aspirin-like substances like almonds, apples, berries, grapes, oranges, plums, and tomatoes for approximately two hours prior to Ototoning; and

5) Watch for medications that may produce tinnitus as a side effect. Not enough information is available to determine if medications can impact otoacoustic emissions.

Steps for Ototoning

These steps are listed in the order for the process but will be followed with more explanation.

1) Exercise for 5-10 minutes before starting;
2) Drink at least 4 oz of water;
3) Find a quiet place without too many sound distractions;

4) Sit in an appropriate chair in the correct Listening Posture. Hands placed on the knees or overlapping together in your lap is best;

5) Breathe deeply a few times to get centered and focused as well as relax the body;

6) Listen for your spontaneous otoacoustic emission. Block out the other extraneous sounds you may be hearing;

7) Try to tonally match your voice with the sound you are hearing;

8) Once identified, close your eyes and make the sound by humming the sound;

9) Produce the sound softly but begin to produce the sound through the bones of the body, towards the back of the throat;

10) Breathe deeply but regularly. Do not over-breathe. Hum the sound on your exhale breath stream. Feel the sound as you make it;

11) Begin to visualize color. What color(s) do you see? Continue humming;

12) Subtly begin to rotate your torso while remaining in the Listening Posture while producing the sound;

13) Continue producing the sound for at least 10-15 minutes, but as long as you want beyond that;

14) Continue supporting the sound throughout the body. Open up more resonating cavities as the sound begins to take effect. Open up the back of the throat into the opening to the ear, so that you feel the bone resonating sound more magnified. Feel the sound vibrating up and down the spine;

15) When you want to stop the sound, when you feel you have had enough for the day, gradually fade the sound out and stop making the sound. Then sit in the silence feeling the reverberations of your Ototoning session pass through and out your body. Stay in the silence of this sound for at least 5-10 minutes if possible; and

16) Experience the change!

Steps and Explanations

1) Exercise for 5-10 minutes before starting.

Exercise for 5-10 minutes before Ototoning to enhance the overall response. Begin by jumping on the trampoline if possible or just jump in place. This can be followed by a series of reverse jumping jacks. Regular jumping jacks prevent your energies from crossing the brain to the other side of the body, so reverse ones allow for the flow of energy. In reverse jumping jacks, stand with your feet spread apart and hands at your sides. Then bring your feet together while your hands go over your head for the jump. And then back to the standing position. An easier activity to do which integrates the energies is a cross crawl. Simply stand in place with the right arm and left leg raised simultaneously and then raise the opposite pair. As you practice, swing your arm across the midline of the body.[19] These exercises begin to activate the semi-circular canals, the vestibule, and the connections to the brain. They begin to set the body in motion.

2) Drink at least 4 oz of water.

Drink water before starting. When beginning to Ototone the body needs to be in the best possible condition for response. Hydration of your body is critical for maximum enhancement of your energy flow. Muscles and connective tissue are comprised of proteins, water, and ions. Movement along a protein metabolism happens like a pogo stick jumping on a line. When moist, this happens in about one ten-millionth of a second, whereas when dry it could require a millionth of a second. The smallest change in hydration can cause a 500 fold increase in the rate of movement. Because Ototoning sets a vibration in place throughout the connective tissue, it is important to keep it hydrated.[20]

3) Find a quiet place without too many sound distractions.

As has been stressed from the first words of this book, silence is golden. It is within the silence around us that we are able to pick up on the important sounds of our life. Silence provides the opportunity to pause for living, silence allows us to recuperate from the constant bombardment of the extraneous sounds around us, and silence allows us to tune into ourselves. It may be impossible to find a totally quiet space, but find one with the least sound distractions. Can you find peace in that space? Do you feel a sense of quietude in this space? Can you feel at one with yourself? The most important sound to stay away from is human vocal sound.

4) Sit in an appropriate chair in the correct Listening Posture. Hands placed on the knees or overlapping together in your lap is best.

Before attempting Ototoning, the body must be in a supportive position to receive the most benefit. Dr. Alfred Tomatis described this posture as The Listening Posture* and is incorporated into all Listening Training Programs (as described within The Tree of Sound Enhancement Therapy). One must sit comfortably on a stool or chair with their feet flat on the floor. Their back is straight and shoulders are back. The back should not be touching anything like the back of the chair or a pillow. The head's position is slightly forward with the chin tilted downwards. The apex of the head should be the highest part of the body. Hands should be placed gently on the knees. With Ototoning, the hands can be palms upwards on the knees or hands can be cupped in the lap. After a few sessions, you will know which works best for you. The hands on the knees provide for a circulation through bone and skin sound conduction of the produced sound, but some people prefer cupping the palms upwards to connect to the external sound world at the same time as internal toning.

5) Breathe deeply a few times to get centered and focused, as well as relax the body.

The Ototoner must breathe deeply first before making a sound. Breathe by opening the back of the throat and taking in breath through your nose. Begin to feel like the insides of the back of the throat are pushing outwards to and through its walls, almost as though a balloon was expanded in the back of the throat and supporting the wall expansion. Sometimes an internal yawn will help expand this area. Opening this area will help distribute the eventual humming sound

---

* More information about The Listening Posture can be found in *Sound Bodies through Sound Therapy* by Dorinne S. Davis, pp. 123-126 and 221-223.

more directly through bone conduction. The sound made often sounds like you are speaking in a closed reverberant room. It is important to breathe in a relaxed manner because this will support the ability to produce more and longer sounds on one breath stream without putting stress on the muscles of the mouth. It is important to keep these muscles relaxed because stress will dampen or block the sensation of sound emitted and received that the body wants and needs.

6) Listen for your spontaneous otoacoustic emission. Block out the other extraneous sounds you may be hearing.

There are a number of ways to accomplish being able to hear your spontaneous otoacoustic emission. You may not have known this sound existed before, so listening for it may take significant practice. For those who are capable of understanding the subtleties of their body, the finding of this sound may be easier.

When mystics are searching for their Anahata sound, students are taught to close their ears with their thumbs and listen for the minute internal sounds of the ear. When in the right mode, this sound should make them deaf to all external sounds. The eyes are closed. Initially the person will hear all sounds but over time these sounds become less noticeable. One sound will stand out. The mind concentrates on this one sound and fixes itself so much on that sound that the mind becomes absorbed in it. External sounds become non-essential. In effect, this is very much the process needed to find our most significant spontaneous otoacoustic emission for Ototoning.

Another approach is easier for those who do not know what to listen for initially. We use a device called a Toobaloo which was developed to support reading, but in effect is incorporating the Voice-Ear-Brain Connection because the voice produces what the ear hears (The Tomatis Effect), and this effect helps improve reading skills. However, I use the Toobaloo as connected with The Davis Addendum to the Tomatis Effect meaning that from this same device we can determine the emission of the ear that will support the voice.*

---

* If you'd like to receive a Toobaloo by mail, please email info@thedaviscenter.com and request one. The Toobaloo is free but you will be responsible for the shipping costs.

The Toobaloo, a long tube with expansions on the ends in the shape of an elongated C, is placed next to the ear on one end and the other end is placed near the mouth. A sound will be heard in the ear which is the resonant frequency of the tube. However after a period of quiet, that sound will eventually cancel itself out and a softer sound can be picked up. That is the otoacoustic emission you need to tune back into the body.

A third way is somewhat similar to the Anahata way. Find the quiet space, sit in the Listening Posture, and start to listen. Try to reach beyond the equipment sounds or the nature sounds around you. Tune into the silence. Sometimes this takes time to learn how to do. We often are not comfortable in silence perhaps because we don't know what to do with the perceptions of sound that our body is experiencing. But it is within the silence that we can begin to find what our body needs and wants. Is there the barest perception of a sound? Does your body sense a sound? If so, tune into that perception or sense.

A fourth way will be the easy way. I will have a device shortly called the Ototoner that picks up the sound from your ear and emits the sound louder for you to hear and produce. This device will identify the specific sound your body wants. When the body suggests that it wants more than one sound, the most dominant sound will be identified.

7) Try to tonally match your voice with the sound you are hearing.

Once the listener has identified that they perceive a sound from their ear they should quietly produce that sound vocally, and try to tonally match that sound. Some people find it difficult to match pitch with any sound. These people are definite candidates for sound-based therapies. They should consider having the Diagnostic Evaluation for Therapy Protocol (DETP) that determines the sequence of therapies needed to support the development or enhancement of the Voice-Ear-Brain Connection. Any person should be able to match pitch; otherwise there is a disturbance in their Voice-Ear-Brain Connection.

8) Once identified, close your eyes and make the sound by humming the sound.

Sitting quietly with your eyes closed helps you focus better on the production of the sound, the visualized color, and the response to the body. Ototoning needs a spontaneous tone, a prolonged and uninhibited sound that incorporates the whole body. The sound should begin as a humming sound. Don't focus on the voice or sound. Focus on feeling the sound in the body. There should be no attempt to control the sound. Let the sound be freely made.

9) Produce the sound softly, but begin to produce the sound through the bones of the body pushing the sound towards the back of the throat.

The sound should be very soft to start. Vocalizing a soft sound is often more what the body wants. Don't be fooled into thinking the body wants loud sound. The louder sound will only evolve as the body feels better about that sound. Sometime we never get beyond a soft sound. The sound's energy is the important piece, not the intensity of the sound.

Bone conduction sound is very important to the success of Ototoning. This means getting the bones of the body singing while humming the needed sounds. Knowing how to get the bones of the body vibrating to support the tone will help move the vibration faster and more efficiently to the entire body. Many authors, musicians, and toners have said that if you can groan or moan, you can tone. Because Ototoning utilizes a hum first and foremost, the best humming includes the entire body. The sound should not only be made in your upper chest, but along your backside as well. Bone conduction vibration allows the sound to move up and down your spine. So, if you hum, use your bum can now be a new catchy phrase indicating moving the sound up and down the spine to reach the entire body. You need to feel the sound in your backside to know you have reached more than just the upper body.

10) Breathe deeply but regularly. Do not over-breathe. Hum the sound on your exhale breath stream. Feel the sound as you make it.

Consider the yawn approach mentioned in step five above. The humming sound needs to be pushing up against the walls of the oral cavity. The sound then is distributed throughout the body through bone conduction. The humming sound should not be a linguistic sound. The humming should be creating a body sound. Humming is only made on the exhaling breath, so maintaining appropriate breathing is important to the process. Humming is a faster way to get to the core level. Resonating the sound with the yawn approach allows you to feel the sound more dynamically. This humming tone assists you from the inside out and outside in. You can feel this when you use the yawn approach. The hum toning itself provides a cyclical sound pattern.

11) Begin to visualize color. What color(s) do you see? Continue humming.

Colors are associated with the chakras of the body and the notes of the scale. Color complements and integrates sound within our body because color as sound has frequency. My story of Ototoning with the evolution of the colors seen may have as profound an effect on you as it did for me. See what colors evolve as you go through your Ototoning session. Throughout the entire session, continue proper breathing and support the humming with bone conduction vibration of sound.

12) Subtly begin to rotate your torso and remain in the Listening Posture while producing the sound.

Because of the connection between sound and movement, once the sounding begins let the body sway slightly to get a feeling of the pulsation of life, similar to our heartbeat or breathing rhythm. This should not be over-exaggerated. Slight rotations of the body torso while in a proper sitting position provide just the right stimulation for our cells to recycle what we need.

13) Continue producing the sound for at least 10-15 minutes, and as long as you want beyond that.

You should devote at least 20-30 minutes for an Ototoning session. Once the sound has been identified and introduced back to the body, a typical entrainment segment appears to take an average of 10-15 minutes before the effect is felt. Some people may need more than this time frame but the listener will begin to notice when the effect has taken place, such as mine when my headache went away. You can maintain the session as long as the tone being produced is effective. Stop once the desired effect has occurred or when the sound no longer feels right.

14) Continue supporting the sound throughout the body.

While Ototoning, open up more resonating cavities as the sound begins to take effect. Open up the back of the throat into the opening to the ear so that you feel the bone resonating sound more magnified. Feel the sound vibrating up and down the spine. Feel the sound throughout the body.

When using the Toobaloo, once the second tone is identified, reintroduce that sound to the body by Ototoning. After a period of time, the listener will no longer hear that sound within the Toobaloo. This means that the body has entrained to that sound and is supporting it within the body by the voice producing that sound. The body has begun to use and integrate the sound it said it needed. In effect, the second sound which is the spontaneous otoacoustic emission, has been cancelled out; it's no longer needed.

If the sound reverberating in the back of the throat appears to be resonating more to one ear than the other, take the pinna or external portion of the ear not receiving the resonating sound as loudly and pull up and back on that ear until the sound becomes balanced throughout the entire oral cavity. The sound must be reverberating throughout the entire body. The person also needs to feel the sound vibration resonating through their spinal column as well as the oral cavity and larynx.

15) When you want to stop the sound, when you feel you have had enough for the day, gradually fade the sound out and stop making the sound.

As mentioned in step thirteen above, plan on at least 20-30 minutes per session although the Ototoning typically takes effect within 10-15 minutes. Once you stop, sit in the silence feeling the reverberations of the Ototoning session pass through and without your body. Stay in the silence of this sound for at least 5-10 minutes if possible.

16) Experience the change!

Enjoy the renewed energy, the sense of well-being, the change in your perceptions!

Additional Steps Available:

1) For those who want to explore various adaptations to the basic Ototoning session, the introduction of language sounds can be used. Some people want to use vowel sounds or various vowel/consonant combinations. One can do this along known resonating areas of the body. The head corresponds to the 'I', the throat corresponds to the 'E', the chest cavity to 'A', the abdomen to 'O' and the pelvis and lower body to 'U'.[21] Vowels produce the most sound and activate the most breath. The overtones contained within the vowel sounds give the sound its energy and resonance. When consonants are added, the sound and energy are partially blocked depending upon the type of consonant.*

Toning for some incorporates the sounds associated with the seven chakras. Jonathan Goldman describes an exercise as "vowels as mantras"[22] where he takes the participant through various sounds associated with the chakras with feeling and intention. The seven vowels from lowest to highest chakras are 'uh', 'ooo', 'oh', 'ah', 'eye', 'aye', and 'eee'. It isn't coincidence that certain names are given to objects or things. If you take the word, *OtoAcoustic Emission*, the O, A, and E are within the vowel sounds of the seven chakras. These particular vowel sounds help connect the pieces of the Voice-Ear-Brain Connection. The 'oh' or O is the third chakra located at the navel and slightly above. One of its energies is with the mastery of self. Ototoning supports mastery of self by balancing where we need to be as well as function. The 'aye' or A is the sixth chakra located in the forehead between the eyes and slightly above them. It is known as the third eye. It corresponds to spiritual awareness. Ototoning supports connecting with one's soul and then connecting with the external spirit. The 'eee' or E is the seventh chakra and is located at the top of the head. It is associated with complete resonance with spiritual energy. This area controls every

---

* For more information on consonants see *Sound Bodies through Sound Therapy* by Dorinne S. Davis, pp. 108-112.

connection between the mind and body. Ototoning supports the balanced energy between the voice, the ear and the brain, thereby connecting the mind and body with its stimulation. Notice that the throat chakra is not included within these sounds, because once we get to this level, we are beyond the voice. We are connecting self (the navel) to the universe (the top of the head). This method of incorporating vowels as related to the three chakra areas and without including consonant sounds can produce a powerful response for those who feel a need to produce something other than a natural sound.

If instead, we consider the O, A, and E as per the known resonating cavities of the body mentioned previously, the sound of the O is in the abdomen, then moves to the A in the chest, and finally into the throat with the E. The movement of the vibration of these sound tonings is still moving upward through the body and expressed with the voice in the throat. With this scenario, the throat or voice is the expression for the natural sound.

With either the chakra or known resonating cavity, sounds associated with the O, A, or E, support the vibratory response of the whole body and links the Voice-Ear-Brain Connection.

2) Another additional step would be to include the overtones of the sound being produced. Overtones can be produced for either the humming sound or the vowel sounds simply by running your tongue over the cavity of the mouth almost in a wave pattern itself from front to back and back to front. The effect of this movement of the sound through bone conduction is very impactful.

Conclusion

Ototoning has the potential to become one of the most powerful self-enhancement techniques ever. There are many techniques that can accomplish similar responses of positive change, but often only with a device. The device created for Ototoning simply provides the listener with the sound needed. That sound comes from the ear. The voice is going to become the tool to make change. Our bodies provide us with that tool. Humanity has been attempting to incorporate this concept with various techniques like toning, mantras, and chanting, as well as some of the techniques used in yoga for hundreds of years. However, it has only been with the introduction of The Davis Addendum to the Tomatis Effect that the final piece of the puzzle has fit into place.

The Davis Model of Sound Intervention evolved through many steps: The Tree of Sound Enhancement Therapy, The Diagnostic Evaluation for Therapy Protocol (DEPT), The Davis Addendum to the Tomatis Effect, training in twenty different sound-based therapies, defining the differences between sound healing, sound therapy, and sound-based therapy, and eventually Ototoning. The Diagnostic Evaluation for Therapy Protocol (DEPT) is the starting place and determines if a sound-based therapy is needed to bring the person into a maximum receptive and/or expressive position to benefit from Ototoning, which is the final piece. A part of this process supports the evolution of a positive solid ego. This often begins to evolve with the use of sound-based therapies as utilized within The Davis Model of Sound Intervention.

If each one of us learns to listen for our own spontaneous otoacoustic emission, and then produce the sound through our bone structure, we are supporting natural self change. The natural self-change can then bring our consciousness to a higher level from which we will be better able to function. Each person can learn to explore this new technique for sound healing called Ototoning.

Sound Notes
Chapter Seven, Summary Statements

- It is within silence that we can connect to our own song.

- Toning, mantras, and chanting have been used as ways to connect the mind/body and maintain wellness for many years.

- Ototoning uses the spontaneous sound of a hum produced vibrationally through the entire body with bone conduction vibration matched to the sound or tone emitted from the ear.

- The sounds coming from the ear as demonstrated by The Davis Addendum to the Tomatis Effect are directly connected to the voice.

- Ototoning addresses the person's needs for their inner self and works towards stabilizing that sense of self or their inner soul.

- Our needed keynote is provided from the emission of the ear.

- Overtones within Ototoning can provide a richer experience.

- An advanced level of Ototoning is when the person can produce the vibrations of their keynote silently in their mind, and yet stimulate the body through bone conduction creating the necessary vibration to self-harmonize the body.

- Ototoning incorporates the connections between the voice, the ear, and the brain.

- Ototoning is a personalized toning session based upon the sound the body tells the person they need.

- In order for the process to work effectively the person should be in balance as demonstrated by the Diagnostic Evaluation for Therapy Protocol (DETP).

CHAPTER 8
A New Paradigm

I have quoted many wonderful people throughout this book – from Pythagoras to Don Campbell. Each in their own way has provided pieces to the unique process which I have called the Voice-Ear-Brain Connection. Each has provided an historic look at why Ototoning is now *the* method for sound healing. This book provides the beginning of a new paradigm, one from which future research should be based.

A new system exists spanning the etheric level and physical level of the body that supports a connection between the voice, the ear, and the brain. This system known as the Voice-Ear-Brain Connection must be in balance for us to feel whole, alive, together and meaningful. The voice and ear are frequency sources for obtaining information about the body. They are also the sound sources and together they form a dynamic system. The ear provides the information about what the body needs and the voice displays how the body is functioning. The voice produces what the ear hears and the ear emits the stressors identified within the voice so the ear supports a cyclical response with the voice's production of sound. Since the ear emits what the body needs and the voice produces what the ear hears, the auditory function of the ear supports The Cycle of Sound. The voice in turn is used to support foundational change to the body. The brain works as the conductor for maintaining the harmony and balance within this system.

From within this system, we can find a way to self-heal. The process comes from sound vibrational energy passing through and outside of our body. How can this be? Everything about us has a frequency or vibration. All matter is vibration. Yet, 'the Word' was first mentioned in the Bible as important to the development of our world. Through the development of language, each of us as individuals has come to be known by our names. Edgar Cayce shared that "names are empowered with a vibrational quality that impels individuals in certain directions or somehow enables them to more readily manifest specific qualities or attributes... The name often embodies the overall vibration and consciousness of the individual." [1]

As humans, what are we? We are 'persons'. The word *person* comes from the Latin word 'persono' meaning through sound. We are people made of sound. The concept of each person having their own resonance frequency or fundamental tone has been discussed by many theorists. James D'Angelo suggested that there is one tone that represents the pure tone of the person.[2] While there is one song print that uniquely identifies a bird by his vocal style and phrasing, so too, it is possible to recognize people by their voices.[3] As with our name's vibration each person also vibrates to the energy of a certain frequency or number.[4]

People talk about receiving good vibes and bad vibes from those around them. In actuality, they are receiving harmonic responses that blend with their own vibrations as good vibes and non-harmonious responses to their body vibrations as bad vibes. When language is used, with good vibes we can express our feelings as feeling alive or great; and with bad vibes we can express feeling vulnerable, distrustful, or anxious. As humans we think with language, but we feel with emotion. Our cells don't have the language per se to communicate, only the communication response that includes interpreting the vibrational sensations or feelings. The response may not be seen, only felt. The energy of the vibration provides the information. We need to tune into these energy signals. If we change the energy, then the response can change.

When Ototoning, one introduces the balancing frequency the body desires. We can maintain our balance with this certain frequency. This can be considered our *keynote*. This keynote in other cultures has been called many names: Sacred Sound by the Chinese, 'the Word' by

Christians, or words such as Aum, cosmic sound, silent sound, or soundless sound. Basically, they all refer to the sound we hear emitted by our ear during silence.

Hazrat Inayat Khan writes: "Every pitch that is a natural pitch of the voice will be a source of a person's own healing as well as that of others when he sings a note of that pitch. But the person who has found the keynote of his own voice has found the key of his own life."[5] This correct natural pitch keynote can only be found accurately by the emission of the ear telling the body what it needs.

We can either Ototone the note as we personally hear it or once the device is created that will provide the note, use that note for self-healing and maintenance. Can these tones change? Yes, they typically will change. Some people need the same sound for days, weeks, and perhaps months until their body self-corrects, but for others, they may need two, three, or more different tones per day depending upon the time of day, what they have eaten, their location, what they have just experienced, and other variables.

Edgar Cayce was asked the question, "What is the note of the musical scale to which I vibrate?" Reading 2072-10 gives his response:

"Thus, as to the note of thy body – is there always the response to just one?... there are certain notes to which there is a response... It is dependent upon the tuning – whether with the infinite or with self, or with worldly wisdom... The tone, then – find it in thyself... This, then, is indeed the way of harmony, the way of the pitch, and the way of the tone. It is best sounded by what it arouses in thee – where, when, and under what circumstance."[6]

Cayce provided that it is important to tone what the body needs. Although toning can be vocalized if desired, it must be at the correct frequency for the body to maximize self-healing. One needs to first identify what the body needs from the output of the ear and then give the body back that sound to bring it in balance.

Once when Cayce was advising about a treatment program for an individual he suggested in reading 3386-1, "Sing a lot about the work –

in everything the body does. Hum, sing – to self; not to be heard by others but to be heard by self."[7] Much of what Edgar Cayce related during his readings involved the vibrations of the body, but with this response he brought in two other key ingredients for making supportive change within the Voice-Ear-Brain Connection: use of the voice and bone conduction humming. He suggested singing which would support self-healing and using humming which would incorporate the bone conduction responses.

Deepak Chopra writes:

> "It is no accident that the syllable Om sounds like the English 'hum'; when the rishis tuned in to the sound of the universe, they actually heard it as a cosmic hum. If you were enlightened, you would be able to hear the vibration that is your own signature; for instance, you could 'hear' your DNA as a specific frequency vibrating in your awareness. Likewise, each neuropeptide would grow out of a sound, as would every other chemical.
>
> Starting with DNA, the whole body unfolds into many levels, and at each one the sutra, or sequence of sound, comes first. Therefore, putting a primordial sound back into the body is like reminding it of what station it should be tuned in to. On that basis, Ayurveda does not treat the body as a lump of matter but as a web of sutras."[8]

What the Voice-Ear-Brain Connection and the important connections with our otoacoustic emissions demonstrate, is that the signature vibration that the rishis heard and used, and the primordial sound put back into the body, is what the Ototoning concept is suggesting but with the exact sound that the body tells us it needs. We know the exact frequency to tune into; we just need to listen for it. Ototoning addresses what the person needs at this moment in time.

The energy of the world comes to you when you need it. Much like the book *The Secret* encouraged people to put positive statements 'out there' and positive responses will happen, so too can the needed energy(s) or connection happen at the right times. Sometimes connections occur when you may not be ready, but the time to make the connection is present within your body's energy, so it happens.

Music therapy started after World War II and although this approach may be considered alternative the results are being verified repeatedly with research. Hospitals incorporate music during surgery, in the waiting room, and even in the patient's room. Some hospitals have designed entire wings incorporating the impact of sound on visitors, patients and staff. Physicians have seen patients healing faster when music has been used.

Connections between the voice and the ear have also been identified as impacting various populations. For example, Duke University Medical Center recently reported that hearing and vocal problems are linked among the elderly more than previously thought. The ear and voice together impact the elderly's communication skills and overall well-being. As a result, depression and social isolation are possible with this age group. In the study released, Dr. Seth Cohen suggested a causal relationship between hearing issues and vocal issues. "When people have trouble hearing, they strain their voices to hear themselves. Likewise, people may strain their voices if their communication partners can't hear."[9] He further suggested that when a person has difficulty with either the voice or the ear that the other modality also be checked for weakness.

Researchers are finding links between music, sound, the voice and ear, hearing and the brain, and biological cellular responses. The time is ideal for setting in place the foundations for how sound can best benefit us all. People want harmony in their lives and they want to take charge of their own body and destiny. The integration of sound energy responsiveness into a wellness regimen should begin now.

Wayne Perry used the phrase "voicing the soul" in his book, *Sound Medicine: The Complete Guide to Healing with the Human Voice*.[10] He considered voicing the soul one's birthright and with it our ability to discover our inner overtone. This appears to be closer to what some consider our fundamental tone, our resonant tone, or our keynote, the one from which our whole body resonates. When I think of 'voicing the soul', I look at finding the sound from our silence, which our inner body or soul is telling us it needs. The sounds that we can re-introduce into our bodies by Ototoning allow us to voice what our body and soul needs. This is truly a way of voicing our soul.

Hazrat Inayat Khan in his book *The Music of Life: The Inner Nature and Effects of Sound,* has written:

> "The way in which man can find his own place is to tune his instrument to the keynote of the chord to which he belongs. Sound is the force which groups all things from atoms to worlds. The chording vibration sounds in the innermost being of man and can only be heard in silence. When we go into the inner chamber and shut the door to every sound that comes from the life without, then will the voice of God speak to our soul and we will know the Key Note of our life."[11]

In effect, this is what we are looking for. From the silence of the world, we will find the sound, or keynote, from our body. This sound comes from the otoacoustic emission from the ear. We stabilize the body by reintroducing this sound to the body with our voice, and then the brain knows how to support us better. We search for reinforcement by listening with the ear. By using our Voice-Ear-Brain Connection, we find our inner sound. We find the sound we need within the silence surrounding us. So, silence *is* golden, and I wish you well finding your golden experience.

About the Author

Dorinne Davis, MA, CCC-A, FAAA, RCTC, BARA is the founder and president of The Davis Center, Inc., the world's leading sound therapy center located in Succasunna, NJ. An educational and rehabilitative audiologist with more than 35 years experience, she received her BA in Speech & Hearing and Speech & Drama, and her MA in Audiology/Deaf Education. She is currently a licensed audiologist in New Jersey, New York, and Pennsylvania.

Certified in Speech Correction, Pre-School Education, Speech & Drama, and as a Teacher of the Hard of Hearing, she has studied and is certified or credentialed in over 20 sound-based therapies including the Tomatis Method®, Earobics™, Aural Rehabilitation, and BioAcoustics™ as a research associate. She is also a certified provider for The Listening Program.

Visit The Davis Center's website www.thedaviscenter.com for more information about Ms. Davis' remarkable work with sound-based therapies and updates on new products and services. And visit her other websites, www.cycleofsound.com, and www.ototoner.com for additional updates. Contact her at info@thedaviscenter.com. Ms. Davis will continue to evolve her Davis Model of Sound Intervention℠, and eventually, she will offer the Ototoner™ for sale and train practitioners in her method for use with both the general public and for therapeutic needs.

Watch for her upcoming book series, *Say it with Sound: Hum, Harmonize & Heal.*

Other books by Dorinne Davis:
- *Every Day A Miracle: Success Stories with Sound Therapy (2006)*
- *Sound Bodies through Sound Therapies (2004)*
- *A Parent's Guide to Middle Ear Infections (1994)*
- *Otitis Media Coping with the Effects in the Classroom (1989)*

Source Notes

CHAPTER 1

[1] "Silence Is Golden", Tremeloes lyrics, accessed 2/2/09, http://www.songfacts.com/details.php?id=2114.

[2] "Silence Is Golden", The Phrase Finder, accessed 2/2/09, http://www.phrases.org,uk/meanings/silence-is-golden.html.

[3] E.D. Hirschk, Jr., et al. *The New Dictionary of Cultural Literacy* (Third Edition, Houghton Millflin Company, 2002).

[4] Deepak Chopra, MD, *Quantum Healing: Exploring the Frontiers of Mind/Body Medicine* (New York: Bantam Books, 1990) 141.

[5] Mary Lynn Kittleson, *Sounding the Soul: The Art of Listening* (Switzerland: Daimon, 1996) 228.

[6] Mary Lynn Kittleson, 230.

[7] John Beaulieu, *Music and Sound in the Healing Arts* (Tarrytown, NY: Station Hill Press,1987) 17.

[8] "The Economic Times", *India Times* article accessed 7/16/09. http://economictimes.indiatimes.com/rssarticleshow/msid-3567485,prtpage-1.cms.

[9] Mary Lynn Kittleson, 29-30.

[10] Kevin J. Todeschi, *Edgar Cayce On Vibrations Spirit in Motion* (Virginia Beach, VA: ARE Press, 2007) 6.

[11] Wiley-Blackwell, "Human-generated Sounds May Be Killing Fish" *Science Daily* (3/13/09), accessed 3/17/09, http://www.sciencedaily.com/releases/2009/03/090312093658.htm.

[12] Richard Gerber, MD, *Vibrational Medicine, The #1 Handbook of Subtle-energy Therapies* (Rochester, VT: Bear & Co., 2001) 60.

[13] Richard Gerber, MD, 137.

[14] Kevin J. Todeschi, 22.

[15] Alpha Galileo Foundation Conference comments, accessed 8/13/08, http://www.alphagalileo.org/index.cfm?fuseaction=readrelease&release-id=531469.

[16] Richard Gerber, MD, 115-116.

[17] *Smithsonian Magazine* articles, accessed 7/10/09, http://www.smithsonianmag.com/science nature/Signal_Discovery.html?c-=y&page=1.

[18] Richard Weaver, and Oleg I. Lobkis, "Ultrasonics without a Source: Thermal Fluctuation Correlations at MHz Frequencies", *Physical Review Letters 87, 134301* (2001) accessed 12/14/11, http://prl.aps.org/abstract/PRL/v87/i13/e134301.

[19] *Science Daily* articles, accessed 7/10/09, http://www.sciencedaily.com/releases/2007/03/070307075703.htm.

CHAPTER 2

[1] Dorinne Davis, *Sound Bodies through Sound Therapy* (Landing, NJ: Kalco Publishing, 2004) 121.

[2] Dorinne Davis, and Sharry Edwards, "BioAcoustic Voiceprint Frequencies and Otoacoustic Emissions", *American Academy of Audiology Annual Convention* (April 2002).

[3] Mary Lynn Kittleson, 12.

[4] Charles Berlin, PhD, Editor, *Otoacoustic Emissions: Basic Science and Clinical Applications* (San Diego, CA: Singular Publishing Group,1998) 22.

[5] A. Lamprecht-Dinnesen, et al. "Effects of Age, Gender and Ear Side on SOAE Parameters in Infancy and Childhood", *Audiology Neurotology* Vol 3 (1998): 386-401.

[6] M. A. Swabey, et al. "Using Otoacoustic emissions as a Biometric", *Biometric Authentication* (Springer, 2004) 600-606. (Lecture Notes in Computer Science, 3072) http://eprints.soton.ac.uk/37942/ 6/25/09

[7] James W. Hall, III, *Handbook of Otoacoustic Emissions* (San Diego, CA: Singular Publishing Group, 2000) 12.

[8] Charles Berlin, PhD, Editor, 62.

[9] http://speech-language-pathology-audiology.advanceweb.com/ Article/Developing-Ear-Cells-Create-Noise.aspx 7/16/09.

[10] http://www.nidcd.nih.gov/news/releases/09/03_16_09.htm 7/15/09.

[11] Dorinne Davis, 30.

[12] James L. Oschman, PhD, Ed., and Leon Chaitow, ND, DO, "What is Healing Energy?: The scientific basis of energy medicine", A series of articles published in the *Journal of Bodywork and Movement Therapies* (New York: Churchill Livingstone, 1997-1998) 244.

[13] Hazrat Inayat Khan, *The Music of Life: The Inner Nature and Effects of Sound* (New Lebanon, NY: Omega Publications, 2005) 232-233.

[14] Laurel Elizabeth Keyes, *Toning: The Creative Power of the Voice* (Marina Del Ray, CA: DeVorss Publications, 1997) 2.

[15] John Beaulieu, 53.

[16] Fabien Maman, *The Role of Music in the Twenty-first Century* (Malibu, CA: Tama-Do Press, 2006)12-13.

[17] http://www.sciencedaily.com/releases/2008/08/080806140209.htm 8/8/08.

[18] http://women.timesonline.co.uk/tol/life_and_style/women/diet_and_ fitness/article6336553.ece 6/24/09.

[19] http://www.alphagalileo.org/index.cfm?fuseaction=readrelease &releaseid=532586 4/27/09.

[20] Mark D. Rapport, et al. "Hyperactivity in boys with Attention-Deficit/ Hyperactivity Disorder (ADHD): A Ubiquitous Core Symptom or

Manifestation of Working Memory Deficits?" *Journal of Abnormal child Psychology* Vol 37, No 4 (May 2009) 521-534.

[21] http://www.usatoday.com/news/health/2007-10-29-exercise-brains_ N.htm 6/25/09.

[22] Bruce H. Lipton, PhD, *The Wisdom of Your Cells: How Your Beliefs Control Your Biology* (Boulder, CO: Sounds True, 2006) Disc 3.

[23] Julia Schnebly-Black, PhD and Stephen, F. Moore, PhD, *The Rhythm Inside: Connecting Body, Mind, and Spirit through Music* (Van Nuys, CA: Alfred Publishing, 2003) 29.

[24] J.C. Birnholz, and B. R. Benacerraf, "The Development of Human Fetal Hearing" *Science*, Vol 222 (Nov 4, 1983) 516-18.

[25] James L. Oschman, and Nora H. Oschman, *Readings on the Scientific Basis of Bodywork,Energetic, and Movement Therapies* (Dover, NH: N.O.R.A.,1997) M10.

[26] Julia Schnebly-Black, PhD and Stephen F. Moore, PhD, 74.

[27] Emile Jaques-Dalcroze, Trans. by Frederick Rochwell, *Eurhythmics, Art, and Education* (New York: Arno Press, 1976) 65.

[28] Julia Schnebly-Black, PhD and Stephen F. Moore, 78.

CHAPTER 3

[1] James L. Oschman, and Nora H. Oschman, A2.

[2] Kevin J. Todeschi, ix.

[3] Hazrat Inayat Khan, *The Music of Life*, 6.

[4] Hazrat Inayat Khan, *The Music of Life*, 39.

[5] Deepak Chopra, MD, 92.

[6] Deepak, Chopra, MD, 135.

[7] Deepak Chopra, MD, 104.

[8] Oksana Trushina, et al. "The Monitoring of Dirty Electricity in a Secondary School in Kazan, Republic of Tatarstan, Russia", *Fresenius Environmental Bulletin*, Vol 18, Vol. 6 (Feb 2009) 1011-1013.

[9] Mark Rider, *The Rhythmic Language of Health and* Disease (St. Louis, MO: MMB Music, 1997) 2.

[10] Deepak Chopra, MD, 122.

[11] Deepak Deepak, MD, 123.

[12] Bruce H. Lipton, PhD, Disc 1.

[13] Bruce H. Lipton, PhD, Disc 1.

[14] Kevin J. Todeschi, 81.

[15] Deena Zalkind Spear, *Ears of the* Angel (Carlsbad, CA: Hay House, 2003) ix.

[16] Gregg Braden, *The Divine Matrix: Bridging Time, Space, Miracles, and* Belief (Carlsbad, CA: Hay House, 2007) 16.

[17] John Beaulieu, 25-27.

[18] Don G. Campbell, *The Roar of Silence: Healing Powers of Breath, Tone & Music* (Wheaton, IL: Theosophical Publishing House, 1989) 85.

[19] Richard Gerber, MD, 241-242.

[20] Deena Zalkind Spear, 90.

[21] James L. Oschman and Nora H. Oschman, C3.

[22] Donna Eden with Davis Feinstein, PhD, *Energy Medicine: Balancing Your Body's Energies for Optimal Health, Joy, and Vitality* (New York: Penguin Group, 2008) 23.

[23] Richard Gilbert, PhD, *Egyptian and European Energy Work: Reclaiming the Ancient Science of Spiritual Vibration* (Asheville, NC: Vesica, 2005) 81.

[24] http://www.smartmoney.com/investing/Stocks/Can-Music-Predict-the-Stock-Markets-Volatility/ 4/13/09.

[25] R. Leichtman, *Nikola Tesla Returns* (Columbus, OH: Ariel Press, 1980) 41-43.

[26] Richard Gilbert, PhD, 17.

[27] Richard Gilbert, PhD, 48.

[28] Richard Gerber, MD, 126

[29] Richard Gerber, MD, 131

[30] Richard Gerber, MD, 119-155.

[31] Fabien Maman, *Raising Human Frequencies: the Way of Chi and the Subtle Bodies* (Boulder, CO: Tama-Do Press,1997) 7.

[32] Gerber, Richard, MD, 119-155.

[33] Gerber, Richard, MD, 119-155.

[34] Gerber, Richard, MD, 155-163.

[35] Gerber, Richard, MD, 162-163.

[36] Gerber, Richard, MD, 189.

[37] Dalai Lama, *The Universe in a Single Atom: The Convergence of Science & Spirituality,* (New York: Morgan Road Books, 2005) 64.

[38] Deepak Chopra, MD, 235-236.

[39] http://catarina.udlap.mx/u_dl_a/tales/documentos/lmu/aguilar_v_a/capitulo0.pdf 3/27/09.

[40] Olivea Dewhurst-Maddock, *The Book of Sound Therapy: Heal Yourself with Music and Voice* (New York: Simon & Schuster, 1993) 51.

[41] http://www.newfrontier.com/1/peru795.htm 3/27/09.

[42] Don G. Campbell, *Sound Spirit: Pathway to Faith* (Carlsbad, CA: Hay House, 2008) xv.

[43] http://en.wikipedia.org/wiki/Boethius 2/2/12.

[44] Mary Lynn Kittelson, 46.

CHAPTER 4

[1] *Webster's New Collegiate Dictionary* (Springfield, MA: G & C Merriam, 1961) 808.

[2] Don Campbell, *Sound Spirit*, 11.

[3] Olivea Dewhurst-Maddock, 93-96.

[4] Don G. Campbell, *The Harmony of Health* (Carlsbad, CA: Hay House, 2006) 57.

[5] http://med.stanford.edu/news_releases/2009/april/brain-waves.html 6/22/2009.

[6] John Beaulieu, 81.

[7] Hazrat Inayat Khan, *The Music of Life*, 10.

[8] http://www.sciencedaily.com/releases/2008/12/081201081710.htm 12/15/08.

[9] Richard Gerber, MD, 208-209.

[10] James L. Oschman and Nora H. Oschman, 11.

[11] James L. Oschman and Nora H. Oschman, J4.

[12] Kevin J. Todeschi, xi.

[13] Kevin J. Todeschi, xv.

[14] Kevin J. Todeschi, 115.

[15] Joyce Whitely Hawkes, PhD, *Cell-Level Healing: The Bridge from Soul to Cell* (New York: Atria Books, 2006) 12.

[16] James L. Oschman, PhD, Ed., and Leon Chaitow, ND, DO, 246.

[17] Candace B. Pert, PhD, *Molecules of Emotion: The Science Behind Mind-Body Medicine* (New York, NY: Touchstone, 1999) 24.

[18] Candace B. Pert, PhD, 148.

[19] Joyce Whitely Hawkes, PhD, 14.

[20] Deepak Chopra, MD, 38.

[29] Candace B. Pert, PhD, 262.

[22] Mark Rider, 38-39.

[23] Mark Rider, 33.

[24] Fabien Maman, *The Role of Music in the Twenty-first Century*, 15-17.

[25] Fabien Maman, *The Role of Music in the Twenty-first Century*, 16.

[26] Dorinne Davis, 238.

[27] James L. Oschman, PhD, Ed, and Leon Chaitow, ND, DO, 121.

[28] Candace B. Pert, PhD, 312.

[29] Deepak Chopra, MD, 83-87.

[30] Deepak Chopra, MD, 87.

[31] Fabien Maman, *The Role of Music in the Twenty-first Century*, 117.

[32] Deena Zalkind Spear, 88.

[33] Kevin J. Todeschi, 62.

[34] Dorinne Davis, 126.

[35] Kevin J. Todeschi, 66.

[36] Kevin J. Todeschi, 76-79.

[37] Kevin J. Todeschi, 99.

[38] Pierre Sollier, *Listening For Wellness: An Introduction to the Tomatis Method* (Walnut Creek, CA: The Mozart Center, 2005) 257.

[39] Fabien Maman, *The Role of Music in the Twenty-first Century*, 111.

[40] Kevin J. Todeschi, 9.

[41] Mark Rider, 27.

[42] Donna Eden with David Feinstein, 33.

CHAPTER 5

[1] http://bridges4kids.org/articles/7-08/Coulter7-7-08.html 6/23/09.

[2] Dorinne Davis, 35.

[3] Deepak Chopra, MD, 127-146.

[4] Wayne Perry, *Sound Medicine: The Complete Guide to Healing with the Human Voice* (Franklin Lakes, NJ: New Page Books, 2007) 57.

[5] Mark Rider, 40.

[6] Fabien Maman, *The Role of Music in the Twenty-first Century,* 81.

[7] Hazrat Inayat Khan, 93.

[8] James D'Angelo, *The Healing Power of the Human Voice* (Rochester, VT: Healing Arts Press, 2005) 5.

[9] Chun Siong Soon, et al., "Unconscious determinants of free decisions in the human brain", *Nature Neuroscience* (May 2008).

[10] Mary Lynn Kittleson, 81-82.

[11] Deena Zalkind Spear, 107.

[12] Deena Zalkind Spear, 108.

[13] Olivea Dewhurst-Maddock, 37.

[14] Olivea Dewhurst-Maddock, 37-40.

[15] Wayne Perry, 31.

[16] Jill Purce, "Sound in Mind and Body", Reprint in *Resurgence* No 115 (March/April 1986) 1.

[17] Darlene Koldenhoven, *Tune Your Voice: Singing and Your Mind's Musical Ear* (Studio City, CA: Time Art Publications, 2007) xi.

[18] Deena Zalkind Spear, 3.

[19] Alfred Tomatis, *The Ear and Language* (Ontario, Canada: Moulin Publishing, 1996) 159.

[20] Julia Schnebly-Black, PhD and Stephen F. Moore, PhD, xv.

[21] Julia Schnebly-Black, PhD and Stephen F. Moore, PhD, 4.

[22] Julia Schnebly-Black, PhD and Stephen F. Moore, PhD, 11.

[23] Deena Zalkind Spear, 122.

[24] Deena Zalkind Spear, 124.

[25] http://www.neuroacoustic.com/biotuning.html 4/23/09.

[26] James D'Angelo, 86.

[27] http://ezinearticles.com/?Meditation-The-Benefits-of-Primordial-Sound-Meditation&id=580772 4/14/09.

[28] Olivea Dewhurst-Maddock, 89.

[29] http://www.vmtusa.com/about.html 7/16/09.

CHAPTER 6

[1] http://www1.ccny.cuny.edu/ci/cru/index.cfm 6/24/09.

[2] James A. Kaltenbach, "Insights on the origins of tinnitus: An overview of recent research", The Hearing Journal Vol 62, No 2 (February 2009) 26-31.

[3] James A. Kaltenbach, 26-31.

[4] James A. Kaltenbach, 26-31.

[5] http://www.medicalnewstoday.com/printerfriendlynews.pho?newsid =97164 4/10/2009.

[6] http://www.medicalnewstoday.com/printerfriendlynews.pho?newsid =97164 4/10/2009.

[7] OHSU study on Tinnitus, report in Oregonian (March 09, 2008).

CHAPTER 7

[1] Don G. Campbell, The Roar of Silence, 75-76.

[2] http://www.organiser.org/dynamic/modules.php?name=Content&pa= showpage&pid=293&page=1 6/8/09.

[3] http://www.hinduism.co.za/anahata.htm 3/27/09.

[4] http://today.msnbc.msn.com/id/30990170/ 6/24/09.

[5] http://www.examiner.com/x-1160-Tampa-Yoga-Examiner~y2009m6d10-Sound-Silence-P2-Being-Quiet 6/24/09.

[6] D. Davis-Kalugin, "Davis Addendum to the Tomatis Effect", 148th Meeting of the Acoustical Society of America, San Diego (November 15-19, 2004) #2pSC14.

[7] Dorinne Davis, 238.

[8] Deepak Chopra, MD, 240.

[9] Richard Gerber, MD, 401.

[10] Mary Lynn Kittleson, 117.

[11] http://www.iammac.org/Mythsandfacts.html 3/27/09.

[12] Don G. Campbell, Creating Inner Harmony: Using Your Voice and Music to Heal (Carlsbad, CA: Hay House, 2007) 9.

[13] Don G. Campbell, Creating Inner Harmony, 47.

[14] Don G. Campbell, Sound Spirit, xvi.

[15] Jonathan Goldman, The 7 Secrets of Sound Healing (Carlsbad, CA: Hay House, 2008) 41.

[16] http://www.ctds.info/tinnitus.html 4/27/09.

[17] A. Shemesh, et al. "Vitamin B12 deficiency in patients with chronic-

tinnitus and noise-induced hearing loss", *American Journal of Otolaryngology*, Vol 14 No 2, (Mar-April 1993) 94-99.

[18] J. Attias, et al. "Oral magnesium intake reduces permanent hearing loss induced by noise exposure", *American Journal of Otolaryngology*, Vol 15 No 1, (Jan-Feb 1994) 26-32.

[19] Donna Eden with David Feinstein, 80-84.

[20] James L. Oschman and Nora H. Oschman, A6-A14.

[21] John Beaulieu, 120.

[22] Jonathan Goldman, 101.

CHAPTER 8

[1] Kevin J. Todeschi, 103.

[2] James D'Angelo, 84.

[3] Olivea Dewhurst-Maddock, 54.

[4] Kevin J. Todeschi, 106.

[5] Hazrat Inayat Khan, 275.

[6] Kevin J. Todeschi, 58.

[7] Kevin J. Todeschi, 97.

[8] Deepak Chopra, MD, 236.

[9] http://www.hearingreview.com/insider/2009-06-04_01.asp 6/24/09.

[10] Wayne Perry, 11.

[11] Hazrat Inayat Khan, 3.

Index

A

academic skills, 27, 28
Acoustical Society of America, 9, 35
acoustic reflex muscle, 9, 26, 118
acupuncture meridian system, 56, 65, 67–68, 71, 106–107
ADHD (attention deficit hyperactivity disorder), 46–47, 89, 120
American Academy of Audiology (AAA), 10, 35
American Speech Language Hearing Association (ASHA), 10
ancient traditions, 45–46, 62–63, 71, 73–74, 108–109, 119, 131–132, 159–160, 190–191
aspirin, 39, 175
ATP receptors, 39–40
audiologists, 8
Auditory Integration Training, 118
    See also Berard Auditory Integration Training (AIT)
auditory processing skills, 27–28
auditory synesthesia, 46
autism, 2, 7, 89, 90, 92, 98, 105, 145, 147–149

B

basal body rhythms, 24–25, 79, 86–91, 106
Beaulieu, John, 3, 44
Beethoven, 100
Bell, John, 58
Berard, Guy, 5, 7, 24, 26
Berard Auditory Integration Training (AIT), 5, 7, 9, 23–24, 26
biblical quotation, 17, 119, 190
BioAcoustics™, 9, 18, 23–24, 35, 41, 93, 100–101, 106, 129, 133–136, 150–151, 167, 174
    See also Edwards, Sharry
BioGeometry, 65
Bio-Tuning®, 130
Bliss, Barbara, 166
body sound energy, overview, 1–21
    author's background, 5–13
    Einstein and, 13
    hearing vs. response to sound, 10–11
    marine mammal research and, 13–14
    reciprocity in sound cycles, 18–19
    silence is golden proverb, 1–2
    soul dimension, 14–15
    sound as energy of universe, 17–19
    vibrational medicine, 14
Boethius, 74
The Book of Sound Therapy (Dewhurst-Maddock), 120
brain. See Voice-Ear-Brain Connection
Buddhism. See dependent origination (Buddhism)
Burr, Harold Saxon, 99–100, 102

C

Cage, John, 3
Campbell, Don, 64, 74, 87, 159, 168, 170, 171
Carlyle, Thomas, 1–2
Cayce, Edgar, 13, 16, 56, 62, 95, 104, 105, 107, 190, 191–192
Center for Neuroacoustic Research, 130
chakras, 68, 115, 131, 163, 168, 184–185
chanting, 12, 17, 46, 117–119, 123–124, 132, 162, 163, 165, 170–171
Chladni, Ernst, 73
Chopra, Deepak, 2, 58, 103, 115–116, 131, 165, 192
Cohen, Seth, 193
coherence, 19, 35, 36, 60, 93–94, 96, 147
color therapy, 167–168
Coulson, Elyse Betz, 43–44
Coulter, Dan, 113–115
Creating Inner Harmony (Campbell), 170
The Cycle of Sound®
    conceptual basis of, 11, 55, 62, 154–155, 189
    as responsiveness, 142
    subtle energy systems, 68, 74
    as system of coherence, 93–94, 152–153
    See also Voice-Ear-Brain Connection℠
Cymatics®, 103

D

Dalcroze, Emile Jacques, 49–50, 128–129
D'Angelo, James, 117–118, 190
The Davis Addendum to the Tomatis Effect
    development of, 9–11, 29
    influence of BioAcoustics™ on, 100–101, 135
    laws of, 31–32, 35, 116, 151–152, 164
    research on otoacoustic emissions, 39, 41
    scope of, 19
    Toobaloo®, 179–180
    The Cycle of Sound®
The Davis Center, 7–9
The Davis Model of Sound Intervention℠, 23–51
    conceptual basis of, 11, 31–33, 141, 145, 154–155
    Diagnostic Evaluation for Therapy Protocol (DETP®), 29–31
    ear physiology and function, 36–42
    entrainment, 88–89
    sound-based therapy, 23–24
    The Tree of Sound Enhancement Therapy®, 24–29
    Voice-Ear-Brain Connection℠, 34–36, 43, 63–64
    See also specific components of Model
dependent origination (Buddhism), 71, 85
Dewhurst-Maddock, Olivea, 120
Diagnostic Evaluation for Therapy Protocol (DETP®), 8, 11, 12–13, 29–31, 32, 97–98, 107, 154–155, 172–173, 180
diet and nutrition, 175

dirty electricity, 58
Distortion Product Otoacoustic Emission Testing, 38
Dousis, Andy, 113–115

E

$E=mc^2$, 13
The Ear and Language (Tomatis), 127
Ears of the Angels (Spear), 63
Eden, Donna, 64–65
Edgar Cayce on Vibrations (Todeschi), 95
Educational Audiology Association, 6
Edwards, Sharry, 9, 18, 28, 35, 41, 44, 133–136, 151
    See also BioAcoustics™
Egyptian and European Energy Work (Gilbert), 67
Einstein, Albert, 13, 63
Electronic Ear (Tomatis Method), 33–34, 127
energy medicine. *See* vibrational medicine, use of term
Energy Medicine (Eden), 64–65
EnListen®, 23-24
entrainment, 66, 88–89, 90, 91, 143, 156, 172
etheric center, 16–17, 63–64, 66–70, 115, 118, 142–144, 166, 189
Eurhythmics method (Dalcroze), 49–50, 128–129
European Science Foundation, 16
evoked otoacoustic emissions. *See* otoacoustic emissions (OAEs)

F

Fast ForWord®, 23–24
4 Minutes and 33 Seconds (Cage), 3
Foxe, John, 145
French Academy of Sciences, 34, 151
Frequency Equivalent™, 28, 100–101, 127–128, 135
Frolich, Herbert, 41, 96, 101
fundamental tones, 85

G

Gerber, Richard, 15, 16, 68–69, 70, 92–93
Gilbert, Richard, 67
Goldman, Jonathan, 172–173, 184

H

harmonics, 66, 72–73, 77, 85, 88, 99, 120–121, 124, 129–132, 147, 151–152, 162–163, 172, 190
The Harmony of Health (Campbell), 87
H'doubler, Margaret N., 47
The Healing Power of the Human Voice (D'Angelo), 117–118
Hearing Sensitivity Test, 29
Hear You Are, Inc., 7
Heidemann, Stephanie, 162
Hippocrates, 98
homeopathy, 102–103
Human BioAcoustics™. *See* BioAcoustics™

I

Interactive Metronome®, 23–24, 49

J

Jung, Carl, 119

K

Karim, Ibrahim, 65, 67
Karpf, Ted, 74, 171
keynote, 190–191
Khan, Hazrat Inayat, 42–43, 56, 191, 194
Kim, Bong Han, 68
Kirchhoff's Principle, 41
Kittelson, Mary Lynn, 2, 3–4
Koldenhaven, Darlene, 125, 126

L

language and speech, 17–18, 26, 28, 48, 113–115, 119, 122, 127, 144
    See also the voice
Law of Similars, 67
Lipton, Bruce, 16, 47, 62
Listening Posture, 50, 126–128, 178, 182
"Listening to Yourself" (Coulter), 113–115
Listening Training Programs, 26–27, 133, 148–149, 173
loudness, defined, 77

M

Maitland, Jeffrey, 55–56
Maman, Fabien, 45, 106, 117
Manners, Guy Peter, 103
mantras, 162–163, 184–185
    See also Ototoning℠; toning
marine mammal research, 13–14
Max Planck Institute for Human Cognitive and Brain Sciences, 118
Maxwell, James Clerk, 63
Maymin, Philip, 66
middle ear infections, 6–7, 26, 30, 154
Molecules of Emotion (Pert), 96–97
Molham, Sophie, 145
Morsella, Ezequiel, 16
The Mozart Effect (Campbell), 168
music
    Boethius on, 74
    effects on body, 86, 104–106
    Eurhythmics method, 49–50, 128
    as medicine, 160–161, 193
    movement and, 45–46
    nerve receptors and, 50
    within own body, 60

physics and, 100
Pythagoras on, 72–73
silence and, 3, 45
sound therapy misconceptions, 75
stock market performance and, 66
See also the voice
Music and Sound in the Healing Arts (Beaulieu), 3, 44
The Music of Life (Khan), 117, 194
The Music of the Elementary Particles (Sternheimer), 100
music therapy, defined, 76

N
neurological system, 33–34, 36, 127, 135, 146
Newton, Isaac, 13, 63
Novalis, 167

O
Oschman, James, 41, 64
Otitis Media (Davis), 6–7
otoacoustic emissions (OAEs), 10, 19, 73, 128
See also spontaneous otoacoustic emissions
Ototoner™ device, 73, 180
Ototoning℠, 159–186
audiation, 125–126
author's experiences with, 167–170
explanation of, 12, 42–45, 155–156
healing effects of, 30, 31, 36, 59
meditation and, 3
need for further research, 18
process of, 173–175
sound properties, 77
sound sensation memory, 49–50
steps for, 175–185
use of own voice, 28–29, 44–45
vocal stability, 136–137
overtones, 77–78, 85, 120, 122, 124, 130, 131–132, 134, 147, 162–163, 172,
184–185

P
Perry, Wayne, 116, 193
Pert, Candice, 96–97, 102
pitch, defined, 77
Plato, 95
Popper, Arthur, 13–14
primordial sound meditation, 131, 160, 164–165, 192
Principle of Nondestructive Coexistence, 15
proprioception, 49
Purce, Jill, 124
Pythagoras, 72

Q

Quantum Healing (Chopra), 103, 115–116
quantum physics, 57–58, 60, 70, 93, 153

R

Rapport, Mark, 46–47
REI, 23–24
The Republic (Plato), 95
resonance, 59–60, 65–67, 74, 92–93, 100–102, 106, 121, 124–126, 131–132, 171, 184–185
resonance therapy, 132
The Rhythmic Language of Health and Disease (Rider), 59, 99, 117
richness of sound, defined, 77–78
Rider, Mark, 59, 99, 117
The Roar of Silence (Campbell), 64, 159
Roberts, Timothy, 92
The Role of Music in the Twenty-First Century (Maman), 45, 117
Rolf, Ida P., 50

S

Sanskrit, 159–160
The 7 Secrets of Sound Healing (Goldman), 172–173
Sheldrake, Rupert, 69
Signature Sound, 101, 134, 135, 164
Signature Symphony of Sound, 32, 66, 99, 102, 135–136, 143, 166, 170
silence, 45
singing, 122–123, 125–126
sonocytology, 103
sound and the body, 83–111
    cell communication, 98–102
    cell patterns and responses, 102–104
    cellular checks/balances, 94–98
    effects of music, 86, 104–106
    effects of sound, 86
    higher frequency responses, 106–109
    natural energy phenomenon, 91–94
    rhythm and energy movement, 86–91
    sound, defined, 85–86
sound-based therapies, overview, 23–24, 32–34, 60–61, 76
Sound Bodies through Sound Therapy (Davis), 7–9, 23–24, 29, 32, 101
sound healing, defined, 75
Sounding the Soul: the Art of Listening (Kittleson), 2, 3–4
Sound Medicine (Perry), 116, 193
Sound Processing Test, 29
sound properties, 76–79
Sound Spirit (Campbell), 74, 171
sound therapy, defined, 76
Spear, Deena Zalkind, 63, 104, 120, 129
speech. *See* language and speech
spontaneous otoacoustic emissions, 38–45, 73, 93–94, 103, 108, 134, 145–146,

150–152, 164, 165–166, 179, 194
Sternheimer, Joel, 100
subtle energy systems, 14–17, 36–38, 42–43, 62–72, 74
Surat Shabd Yoga, 3
systems and basic sound, 55–79
    sound intervention, 72–76
    sound properties, 76–79
    total person and, 55–56
    vibrational patterns of body, 56–62
    Voice-Ear-Brain Connection^SM, 68–72, 79
    whole body subtle system, 62–72
Szent-Gyorgyi, Albert, 101

T
Tesla, Nikola, 66–67
testing. *See specific diagnostic tools*
Thompson, Jeffrey, 130
throat singing, 131–132
Tiller, William, 70
timbre, defined, 77
tinnitus, 40, 145–146, 147, 164, 175
Todeschi, Kevin, 95
Tomatis, Alfred, 9, 27, 33–34, 56–57, 104, 105–106, 116, 126–128, 151–152, 171–172, 178
The Tomatis Effect, 31–32, 34–35, 133, 151–152, 179
Tomatis Listening Test, 29
The Tomatis Method®, 23–24, 126–128, 132–133, 151–152, 171
toning, 12, 17–18, 75, 117–119, 123–124, 162–163, 170–171, 184–185, 191
    See also Ototoning^SM
Toobaloo, 179–180, 183
Transient Evoked Otoacoustic Emission Testing, 38
The Tree of Sound Enhancement Therapy®, 8, 9–10, 120, 154
    academic skills (upper leaves/branches), 27, 28
    auditory processing skills (lower leaves), 27–28
    basal body rhythms (seeds), 24–25, 79, 86–91, 106
    effectiveness of, 74–75
    maintenance/support (head), 28–29, 107
    sense of hearing (root system), 26
    sound-processing skills (trunk), 26–27
    total person and, 55–56, 62, 72
Tune Your Voice (Kodenhaven), 125

V
Vibrational Medicine (Gerber), 15, 16, 68–69, 92–93
vibrational medicine, use of term, 13, 14, 141, 149
vibrational patterns of body, 56–62
vocal scanning, 131
the voice, 113–137
    BioAcoustics™ and, 133–136
    body vibration and, 120–121

connecting to inner voice, 117–119
    as instrument, 136–137
    listening beyond the words, 121–125
    spin off methods, 132–133
    Tomatis on, 126–128
    vocalizing and vocal techniques, 125–126
    whole person response, 115–116
Voice-Ear-Brain Connection^{SM}, 4, 10, 141–156
    body aspect, 149–151
    body movement and, 47–51
    brain aspect, 147–149
    ear aspect, 40–41, 144–147
    evolution of discoveries, 153–156
    five laws of, 34–36, 152
    foundation of, 151–153
    Ototoning^{SM} and, 12, 19, 44–45
    rhythm and, 88, 91, 106
    sound-based therapy modalities, 23–24
    subtle energy system of, 15–17, 42–43
    as system, 142–143
    techniques, 128–132
    voice aspect, 116, 143–144
    wholeness and, 30, 121, 151–153
    See also The Cycle of Sound®; systems and basic sound
Voice Movement Therapy, 132
voice tuning technique (Spear), 129

W
The Wisdom of Your Cells (Lipton), 16, 47, 62

Y
yoga, 3, 18, 74, 108–109, 131

CPSIA information can be obtained at www.ICGtesting.com
Printed in the USA
BVOW04s1736231013

334462BV00007B/55/P